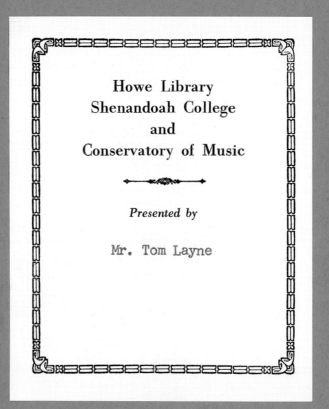

THE MATHEMATICS LABORATORY
Theory to Practice

THE PRINDLE, WEBER & SCHMIDT SERIES IN MATHEMATICS EDUCATION

Consulting Editor:
John Wagner
Michigan State University

THE MATHEMATICS LABORATORY
Theory to Practice

co-authored by

ROBERT E. REYS
University of Missouri

THOMAS R. POST
University of Minnesota

introduction by

ZOLTAN P. DIENES

PRINDLE, WEBER & SCHMIDT, INCORPORATED
Boston, Massachusetts

Library of Congress Catalog Card Number: 72-92803
Printed in the United States of America
SBN 87150-161-9

Second Printing: August, 1974

CONTENTS

An initial thought:

"If God knew what kind of children the schools would require, he would have made them much differently."

Source unknown

INTRODUCTION

As every mathematics teacher or teacher of elementary school children knows, there has been a vast revolution in our attitude to the teaching of mathematics in the past ten years. This change was necessitated for the most part by economic considerations, but it was also stimulated by rivalry with other countries in the power struggle between political blocks. Be that as it may, it is now commonplace to say that the teaching of mathematics is a problem of international importance. In most countries a short-term view was initially taken of the ways and means of tackling this problem. It was found that insufficient training in mathematics was given to those leaving school, and so, the content of the high school mathematics courses was changed so that better prepared candidates could be welcomed at institutions of higher learning. Of course, this orientation towards content did not take into account the obvious fact that, as a radical increase in the number of mathematically trained persons was required, then naturally, a radical change in the situation in which the learning took place became a necessity. It is only in recent years that it has been realized in most countries that not only a content revolution but also a methodology revolution is required. England has been the spearhead of this methodology revolution; in that country it began long before the content revolution and has now culminated in what is known as the Nuffield Project, itself based on several decades of pioneering work in experiments such as the Leicestershire one.

The authors of this book question the premise that mathematics can be learned simply by learning the language in which it is expressed. This language is the system of symbols used to convey mathematical ideas, usually from one mathematician to another. The conveying of such information involves the thorough understanding of the many links in the content conveyed. It is possible, however, to learn a symbol system and even its utilization in some restricted number of cases without knowing what content the symbol system is, in fact, attempting to convey. This is still the situation in the vast majority of mathematical learning situations, even at the university level. It is a very uncomfortable truth to face, that in fact, extremely few of us know any mathematics and extremely few of us are learning any. We are beginning to realize now that the reason for this is not that we have tried to teach the wrong content, but that we have tried to teach at all. To "teach" a concept is essentially impossible. It is only possible to *learn* a concept. The proverbial horse can be taken to the water, but he has to decide to do the drinking. In other words, the old fashioned theory of education, which assumes the existence of a funnel through which knowledge is poured into the brain, has to be abandoned. The laboratory approach here advocated is based on a vast body of theoretical research into learning by persons such as Piaget, Gagné and myself, who have experimented in actual situations in order to build up a theoretical background out of which practical suggestions can be concluded. Such suggestions form the backbone of the present book. The laboratory situation is, in effect, an answer to the *teaching versus learning* controversy, if it can be called one. In order to make learning virtually certain, we need to place the child in a situation which will motivate him. A highly motivating situation is like a suitably high octane gasoline that we put into a motor vehicle. If we put a low octane gasoline in a car which is supposed to take high octane, it will not perform very well, or may not even start. We need to use a different octane learning material now, and the practical problem is to find out just how such material can be constructed, test its use, and do research equivalent to that done by motor car and gasoline manufacturers in order to sell their products. The difference between the two cases is simply that there is competition in the motor car and in the gasoline industry and there is none in education. Because of the simple reason that they have no competition, teachers and education authorities are permitted to work at a rate of inefficiency which would render most businesses bankrupt within twenty four hours. It is probably true to say that any mathematical reform which does not aim to increase learning efficiency by several hundred percent is not worth considering.

Every teacher of mathematics knows that the first thing to do when he gets a class in the beginning of the year is to review every-thing, because he is certain that everything will have been forgotten. He will usually blame the previous teacher for inefficiency but, in turn, he will be blamed for the same inefficiency by the next teacher who takes on the same class. A favourite subject for review is frac-tions; every year we review fractions and every year we still need to review fractions, because the children have not understood. Very few job analyses have been done about the hierarchies of learning, such as those suggested by Gagné and others, to determine what the prerequisites are for operating in situations requiring fractions. Such information could be extremely useful to both curriculum develop-ers and classroom teachers. Clearly, continued widespread efforts are needed if we are to find more effective ways by which children can learn mathematics.

There is now a small but quite widespread international team, the members of which are trying in the small ways open to them to remedy this situation. This team has voluntarily banded to-gether and organized itself as the International Study Group for Mathematics Learning; it has a number of constituent groups in about 20 different countries in the world. The problems tackled by members of this group fall into three main categories:

1.   Psychological research into learning. It is necessary for us to know more about the actual process by which learning takes place in order to effectively apply new mathematical approaches. This kind of theoretical research is being undertaken by certain members of the group. The novel aspect of these research projects is that they are not being confined to the laboratory, but are being compared in the same project, or in some other sister projects, with immediate classroom use and application. For example, the study of the viability of the multiple-embodiment principle presents a challenging psycho-logical problem in experimentation. So does the problem of general-ization and of symbolization, which involve the twin but opposite activities of encoding and decoding.

2.   At the same time we are trying the in's and out's of the psycho-dynamics of the above processes, we are trying to get prac-tical information concerning their working in the classroom and then collate this information. This practical work provides valuable feed-back for the theory builders who will have other hypotheses to test. Yearly meetings are held by the theoretical research oriented members as well as by the members of the more practically oriented projects. The latter are usually conducted in the form of seminars and workshops where teachers from different countries are invited to participate, give demonstration lessons, discuss work with mate-rials, and generally learn from each other. This kind of international and national cooperation between different research teams and practical working teams in the field should become even more wide-spread in the future.

3.   Solving problems of curriculum construction and the inevitable concomitant problems of teacher training form another important part of the research programs. Much more mutual aid at all levels will be needed before we are "out of the woods."

It is high time that the second round of reform in the mathematics learning situations should begin. This second round will concern itself mainly with methodology, as the content areas have already been well defined. In fact, probably as the methodology reform progresses and becomes more and more successful, it may become evident that the content considerations will become correspondingly less and less important. After all, we do not know just what mathematics a particular child will need in forty years time, when he is working on some project yet completely unknown to us. It is possible that the mathematics he will need then hasn't yet been invented. We need to educate children to meet the unexpected, to be able to be confident in themselves that they can tackle a problem using a highly complex and refined structural apparatus. In other words, we have to teach him how to *learn structure*. You learn most things by doing them. You learn to ski by skiing, you learn to ride a bicycle by riding a bicycle. So you have to learn to learn structures by learning structures. When a child has sufficient practice in learning complex and refined structures he will be able to learn other complex and highly refined structures more easily. We use the multiple embodiment principle in order to give him practice in abstraction, and we use the mathematical variability principle in order to give him practice in generalization. Then of course, he also has to be given practice in coding and decoding, that is, in writing a symbolic language as well as reading it. Sets of such skills and some of their sub-skills will probably soon replace curricula. Hopefully, we shall be able to identify some of the main requirements of intellectual skills that underlie the learning of most of these complex mathematical structures. Using such principles as the multiple embodiment and the mathematical variability principle, involves the use of a vast amount of concrete material.

Chapters 6 and 7 of this book provide noteworthy examples of how these principles can be used to design well-defined sets of student learning activities which utilize a wide variety of learning materials and yet focus on a central theme or mathematical topic. The series/parallel model illustrated in these chapters is one which I feel has considerable potential, for it defines a process whereby clusters of related learning activities can be designed and structured so as to promote both abstraction and generalization. Teachers need to acquire skill in both selecting such activities and materials and in providing children with the opportunity of making good use of them in their learning. This is the main purpose of this book. It is suggested that the provision of materials and consequent individual or small group learning situations represent a considerable departure from the classical idea of the teachers' role. This role will have to be radically redefined and the underlying philosophy rethought. The very look of the classroom will need to undergo a considerable change. The organization of activities, the noise level permissible will be different: practically every "sacred cow," so far carefully sheltered from destruction, will have to be given up and replaced by more modern and more effective methods of learning.

I have great pleasure in introducing this book. I earnestly hope that a wider acceptance of the principles and practices contained in its pages will do much to liberate children's minds from the intellectual slavery to which they have been condemned by traditional classroom practices.

Zoltan P. Dienes, Directeur
Centre De Recherches en
Psycho-Mathématique
Université de Sherbrooke
Sherbrooke, Québec, Canada

PREFACE

The modern classroom teacher is charged with responsibilities, the magnitude of which can only be described as awesome. Not only are teachers in large measure responsible for the physical, social, emotional and intellectual development of many children, but they are, at the elementary level, also expected to be knowledgeable about and proficient in teaching eight to ten different subjects. In addition, teachers are strongly urged to individualize, personalize, and humanize each subject for each child. It is doubtful whether any single mortal human being could satisfy such professional expectations even under the best of conditions. It is perhaps foolhardy to expect such performance from several million persons under conditions which are clearly less than optimal.

New roles for both teacher and student must be identified, new content delivery systems must be designed and new emphasis must be placed on the humanistic aspects of the educational process if the expectations outlined above are even to be approximated. We believe that the concept of the mathematics laboratory as developed in this manuscript addresses itself to all of these concerns insofar as they apply to the teaching and learning of mathematics in the school setting. *We have defined the mathematics laboratory as both an approach to learning as well as a physical environment.* Such a definition provides for greater potential classroom utilization of the laboratory concept. Certainly not all schools can afford elaborate materials and equipment, but every classroom teacher could incorporate this approach to learning in his interactions with children.

This book should be of interest to the pedagogue who feels the need for a greater degree of student involvement in the mathematics program, to the person interested in exploring alternative classroom strategies, and to the instructor concerned about making mathematics a more relevant part of the child's everyday world. In the final analysis, this book was written to assist classroom teachers interested in initiating a laboratory approach in their mathematics classroom.

Two major premises influenced the preparation of this manuscript:

1. Teachers should function as facilitators of learning, not as dispensors of information.
2. Teachers tend to teach as they have been taught.

The first premise recognizes the futility of any person assuming the role of primary knowledge source and the second implies that teachers must personally experience alternative instructional strategies if they are to try new approaches in their own classrooms.

The authors believe that many teachers would like to utilize an active approach to the learning of mathematics, but are unsure as to a viable method for going about the business of implementation in the classroom setting. Questions related to this dilemma considered in the book include:

1. What is a mathematics laboratory? Why have one?
2. Is the concept of the mathematics laboratory consistent with what we know about the way children learn?
3. What criteria should be used in selecting and evaluating laboratory activities?
4. Can I, as a classroom teacher, develop high quality laboratory activities for my own classroom? How?
5. How can I integrate a laboratory type approach with the existing mathematics program?
6. Does my role as a classroom teacher change as a result of this approach?

We hope that our responses to these and other important questions will help clarify in the reader's mind the essential components of a relatively new (new in terms of its widespread acceptance and implementation) approach to mathematics instruction. In addition, this book provides practical suggestions so essential to success in the classroom situation.

You may also be interested in other features of this book:

1. This is *not* a book of mathematics laboratory activities. Certainly some activities have been included. However, these lessons were developed primarily to exemplify the kind of curriculum development that we feel is badly needed. Hopefully, these lessons can also serve as prototypes for future lessons.

2. The reader is given some determination in his fate. For example, the chapter "The Mathematics Laboratory in Theoretical Context" provides an "overview" which is followed by a development of learning theory from the "behaviorist and cognitive" positions. The latter part of the chapter is devoted to "Conclusions and Reactions" and "Summary and Implications." We recognize that not every reader is ready for nor interested in a discussion of these theoretical positions at the outset. Therefore, this chapter has been designed so that readers (who wish) may skip these discussions initially, returning to them at a later time when their relevance has been more firmly established. The reader will be referred back to various sections of this chapter throughout the text, for it provides the theoretical context upon which the curriculum development model and related laboratory activities are ultimately based.

3. Common threads weave their way through the chapters. Although the classroom facilities, instructional materials and role of the teacher are discussed in distinct chapters, it is recognized that they are interrelated and in fact inseparable. The discussions within these chapters reflect this position.

4. The ideas and theories discussed are followed by illustrations and pragmatic suggestions. The rationale and development for active learning of mathematics is linked to the actual implementation of these ideas. For example, providing for "multiple embodiment" of a mathematical concept is presented initially in Chapter 3. It also provides the basis for a curriculum development model suggested in Chapter 5 and is further clarified by examples in Chapters 6 and 7. These activities serve to illustrate the theoretical model in specific learning situations.

All of the materials have been used with both elementary and secondary teachers at the University of Missouri and the University of Minnesota. The authors gratefully acknowledge the many students and teachers who have used these materials and provided us with valuable feedback.

We would like to express our appreciation to William M. Fitzgerald (Michigan State University), Carole E. Greenes (Boston University), Douglas Grouws (University of Missouri), Loye Hollis (University of Houston) and James W. Wilson (University of Georgia) for reading and reacting to early drafts of this manuscript. We also extend our gratitude to Jeanne Bursheim (Osseo, Minnesota), Nancy English (St. Louis, Missouri), Robert Jackson (University of Minnesota), Joan Kirkpatrick (University of Edmonton), and Phyllis Mirkin (Minneapolis, Minnesota) for the contributions reflected in the appendices and to Frances Huffman, Pat Rush and Cindy Tatum for developing early drafts of the lessons on numeration systems. In addition, our thanks to Claudia Trautmann for her patience in typing the final draft.

Finally, we are indebted to the genius of Jean Piaget, Jerome Bruner, and Zoltan Dienes. Their writings have caused us to rethink, redefine, and redevelop our own philosophies of mathematics learning. We hope that in some small way this manuscript will help the classroom teacher to make mathematics a more vibrant, a more stimulating, and a more meaningful subject for every student under his influence.

Robert E. Reys
Thomas R. Post

The "revolution in school mathematics," weathered by a decade of feverish innovation, has subsided somewhat in intensity, permitting a more deliberate, unemotional and perhaps a more rational examination of where we have been, where we are now, and where we should be going. The magnitude of the changes undertaken has been great in both mathematical content and its accompanying pedagogical techniques. These changes have necessitated major role modification for both teachers and students. The teacher can no longer be considered the source from which all knowledge emanates, and the student is no longer expected to passively absorb mathematical knowledge by imitating the action of the teacher in repetitious "problem" situations.

The results of recent psychological investigations by men such as Robert Gagne, Jean Piaget and Zoltan Dienes about the ways in which children learn are affecting the way in which pupils are being taught. By and large, the net result of this influence has been the introduction of more and improved materials in the mathematics classroom. Effective teachers consider active pupil involvement in the learning process, not only desirable, but essential if the child is to reach his mathematical potential. No longer can an adequate program in mathematics at any level be completely contained between the covers of a single textbook.

The recent international growth, development, and implementation of learning activities in mathematics classrooms via an informal laboratory approach has been great. Among the projects that have contributed to this development are the Adelaide Mathematics Project (Australia), the Madison Project (United States), and the Nuffield Project (England). In addition to these projects, there are thousands of elementary, junior high, and senior high school teachers who have been actively involved in mathematics laboratories of one kind or another.

Considerable work related to mathematics laboratories has been going on for some time. Several books and numerous pamphlets have appeared recently, discussing the virtues of actively learning mathematics. Yet these publications are dwarfed by the vast number of laboratory activities, problem cards and commercial apparatus currently being produced. Classroom teachers are literally overwhelmed by the sheer magnitude of new materials and the concomitant advertising campaigns encouraging their adaptation. Clearly, continued efforts are needed to attempt to put current innovation into perspective and to provide the classroom teacher with the resources to improve the quality of mathematics instruction. It is toward this end that subsequent chapters are directed.

CHANGING ROLE OF TEACHER AND STUDENT DEMANDS INSTRUCTIONAL ALTERNATIVES

CHANGING WORLD OF EDUCATION. The rapidity with which various components of our social structure are undergoing modification is astounding. The knowledge explosion has had its effect in all areas affecting our daily existence. New developments in mathematics, physics, chemistry, biology, cybernetics, and general technology, to name but a few, have had and will continue to have profound influences on the way in which we live. Innovation has not been restricted to the physical and natural sciences. The fields of economics, medicine, sociology, and psychology have continued to keep pace with the tempo of current developments.

Pedagogy, in general, and mathematical pedagogy, in particular, certainly need to be included in any list of fields in which an intellectual revolution has occurred. Classroom implications of certain ideas in educational psychology (e.g., psychology of authority, drill, and reinforcement) have long been recognized. However, more recent developments in educational psychology have provided educators with more objective criteria on which to base educational decisions concerning the appropriateness of mathematical content and various pedagogical techniques. In short, we know more today about the way children learn mathematics and the general nature of the mathematics they are capable of learning at various stages of cognitive development than we have ever known before. Ironically, we still do not *KNOW* precisely how children learn, but the efforts of researchers are continually providing new evidence to support (and oftentimes refute) various learning theories. Since learning is an individual matter and invariably dependent upon numerous factors— some of which are quite elusive—it is highly unlikely that a comprehensive learning theory which is all things to all people will ever evolve. Suffice it to say that we are continually adding to the existing bank of knowledge in this very complex area. Notable among the current investigators in the general nature of intelligence and learn-

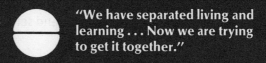

ing are J. P. Guilford, Jerome Bruner; Jean Piaget, Zoltan P. Dienes and Robert Gagne have made significant contributions toward the formulation of theories specifically germane to the area of mathematics learning.

Although much has been written about the "new mathematics," "modern mathematics" or the "revolution in school mathematics," there is no clear consensus whether innovation will continue to predominate. To illustrate these two positions, consider the following statements which were both expressed in 1967. Joseph Hooten said, "The first wave of mathematics education reform has crested in the United States."[1] Robert Davis, on the other hand, declared" . . . the new mathematics revolution has not taken place, but—considering the pressures that are building up—it probably will, possibly within the next ten years. . . ."[2] If the past 15 years have not witnessed a full scale revolution in mathematics, it at least represents a period of considerable turbulence.

The ever increasing rate of general social change, together with the knowledge explosion, have played no small part in the wide scope of curricular innovation taking place in the nation's schools. No end is in sight for the rapid rate of social change or for the pace at which new knowledge is being produced. Under these circumstances it is not possible to predict the precise nature of tomorrow's society or the precise kinds of knowledge, skills, and abilities which will assure today's students a productive role in tomorrow's world. Since the schools are primarily responsible for developing the intellectual tools demanded by society, the nation's educational systems must

---

[1] Joseph A. Hooten, Jr. (Editor). "Proceedings on National Conference on Needed Research in Mathematics Education," *Journal of Research and Development in Education*, 1:1, Fall 1967.

[2] Robert B. Davis, *The Changing Curriculum: Mathematics*, (Washington, D.C.: Association for Supervision and Curriculum Development, 1967), p. 1.

continue to be cognizant and sensitive to the needs of the society which it serves.

TRAINING VS. EDUCATION: CHALLENGE FOR TODAY'S SCHOOL.
The current challenge facing the nation's schools is like no other in history. It must provide experiences for young people that will nurture and develop their capability to assume a productive, contributing role in tomorrow's society. This challenge can be viewed in at least two ways. The first is that educational systems need to produce "problem solving" individuals, where problem solving is defined as "a response to a problematic situation for which the individual involved has no ready-made response pattern." "No ready-made response pattern" is the key phrase here; it implies that individuals must be capable of original and creative types of thought processes in order to solve problems never before encountered. It also implies that individuals must be able to organize and apply existing knowledge and problem solving techniques to new situations.

The second way in which this challenge can be viewed is as a demand that concern for "education" supercede the concern for "training" in the hierarchy of educational objectives. Let us consider some essential differences between education and training, several of which are identified in Figure 1.

TRAINING vs. EDUCATION:

SOME ESSENTIAL DIFFERENCES*

Training	Education
1. Stresses memorization	1. Emphasis on understanding
2. One way communication— from teacher to pupil	2. Two way communication— mutual exchange between teacher and pupil
3. Fosters pupil dependence	3. Encourages pupil independence
4. Major emphasis on past and present situation	4. Preparation for present and future

Figure 1

---

* This suggests a clear dichotomy, where "training" is associated with bad and "education" is associated with good. The point is not to present a strawman argument in favor of education; but rather to remind readers of elements that characterize these types of learning. There are times when training is essential as it often serves as a prerequisite for continuing education. The challenge is to keep training and education in proper perspective.

Training stresses imitation or memorization, and thus is oriented toward maintaining the status quo. Education emphasizes preparing the individual so that he "... is capable of going beyond the cultural ways of his social world, able to innovate, in however a modest way, so that he can create an interior culture of his own."[3] Hence the focus of education is directed toward the future.

Training generally entails one-way communication—from teacher to student. Education encourages and requires a mutual exchange of ideas and feedback, between teacher and student as well as among students. The trained person looks toward other persons for instruction, whereas the educated person is capable of developing increasing independence. The "trained mind" is at home with the familiar situation, but ill at ease when placed in a strange setting. The trained mind may therefore find it extremely uncomfortable or next to impossible to operate effectively in our complex world of change. The "educated mind," on the other hand, is more adaptive in an atmosphere of uncertainty and is more likely to accept and welcome opportunities to function in a constantly changing environment.

The role of the teacher actively engaged in the education of his students is considerably different from that of his colleague who perceives education and training to be synonymous. The "educating teacher:"

a. Carefully selects learning experiences which promote active student involvement,

b. Believes in the intellectual capability of his students and plans his instruction accordingly,

c. Perceives his role as a facilitator of learning rather than the source from which all knowledge emanates,

d. Recognizes the importance of the "why" as well as the "how",

e. Deemphasizes exposition while promoting student discovery,

f. Encourages student hypotheses (guesses) and their subsequent evaluation,

g. Encourages and rewards divergent student responses (creative thinking),

h. Promotes continuous classroom interaction (student to student and teacher to student),

i. Continually promotes genuine problem solving activities, and

j. Recognizes the importance of the process as well as the product of learning.

There are undoubtedly numerous ways for teachers to accomplish many of these methodological objectives. Although this book is concerned with an active learning approach, the authors

---

3   Jerome S. Bruner, "After Dewey What?" *Saturday Review*, p. 59, June 1961.

do not intend to imply that this way of learning is best for all students. It is generally agreed that all children do not learn equally well from the same instructional method.* For some, a good expository presentation is most effective, for others a guided discovery lesson will produce greater gain, while still others may learn best in an independent study situation. *Research has not provided conclusive evidence regarding the efficacy of any single mode of teaching in all situations.*

Even though there is not a *best* instructional method for all teachers, it is easy to get into a "teaching rut." We can become victims of a routine and consequently teach things the same way over and over again. One must carefully avoid the temptation to use only one teaching style, for this can promote a "one right way of teaching" syndrome making the danger of classroom sterility very real.

NEEDS OF TODAY'S SCHOOL.   A cursory examination of current essays on our schools raises serious questions about the existence of a proper educating environment.† Much criticism has been directed toward our schools. Two of the most frequently voiced criticisms are that classes (both elementary and secondary) are generally (i) exposed to a relatively sterile learning environment and (ii) are dominated by traditional teacher-centered activities. The latter generally takes the form of lectures or "show and tell" presentations. The authors do not believe the traditionally managed classroom to be the vehicle or medium through which the educating teacher can most efficiently operate.

A different type of environment is needed by children of all ages to facilitate the effective and efficient learning of mathematical skills and concepts. For the student it must provide sources of remedial and enrichment activities, it must facilitate and encourage exploration, and it must provide a source of manipulative materials with which he can experience mathematical ideas appropriate to his learning and/or ability level.

Furthermore, it must free the teacher to provide the guidance and assistance often required by individual students, it must increase the resource capability of the individual classroom, and it must be capable of intrinsically motivating students to become involved with mathematical ideas. We believe that the mathematics

---

* The implication is clear. Teachers must be able to successfully use many different instructional methods, i.e., teachers must teach! And even more important, teachers must have the wisdom and professional judgment to use methods appropriate for specific students as they study particular topics. This is a lofty goal and presents a continuous challenge to teachers. Remember, however, that successful teaching is not a destination, but rather a journey.

† Books such as *Crisis in the Classroom, Schools Without Failure, Freedom to Learn* and *How Children Fail* provide a candid appraisal of the current situation in many of today's schools.

laboratory is capable of providing that kind of environment. Although not a panacea for all educational ills, the use of a mathematics laboratory represents an instructional approach about which every teacher should be aware.

WHAT IS A MATHEMATICS LABORATORY?

The phrase mathematics laboratory, although difficult to define, has at least two distinct connotations. One is that of an approach to learning mathematics, while the other is that of a place where students can be involved in learning mathematics. These notions encompass the physical aspects of a room with materials and the educational purposes which the laboratory is designed to achieve.

To some teachers a mathematics laboratory may mean materials stored in a cabinet or closet in a classroom. To some it may mean specialized mathematics equipment, games and puzzles. To others it may suggest a classroom that includes a demonstration table, numerous manipulative materials, and an array of different types of apparatus. Although these components are a necessary part of any definition of a mathematics laboratory, such views, in and of themselves, are not sufficient. The concept of the laboratory transcends a mere collection of learning aids. We conclude that each of the above is a narrow and very limited conception of the mathematics laboratory. Although physical materials are important, the manner in which they are used is crucial. Adequate physical resources coupled with sound instructional technique are both essential ingredients of the "mathematics laboratory" considered here.

A child's concept of a mathematics laboratory provides a different and yet interesting perspective. Rasmussen says:

> To the child, the mathematics laboratory is a playroom where things can be counted, moved, rearranged, stacked, measured. . . . It is a room with things to be weighed and instruments to weigh them with, machines with buttons and levers and cranks that count, record and project. A room with books for browsing in . . . books to be written in. A room with objects of many different shapes and sizes to be used for building and comparing."[4]

This description characterizes a mathematics laboratory and suggests an informal climate for learning. The implied objective of the laboratory is to actively involve pupils in the learning process. Such a format for learning is essential for students of all ages.

A mathematics laboratory is much more than physical materials or a physical environment for learning mathematics. In fact the mathematics laboratory itself reaches far beyond materials in a classroom and utilizes any facilities needed (whether in the school, play-

---

[4] Lore Rasmussen, "Countering Cultural Deprivation Via the Elementary Mathematics Laboratory" THE LOW ACHIEVER IN MATHEMATICS, Washington, D.C.: United States Office of Education, 1965. p. 70.

ground or home) together with a spirit of inquiry that must prevail in both pupils and teachers. This spirit of inquiry so fundamental to the *laboratory approach*,* involves students in a "learning by doing" situation. It provides an informal climate for learning that creates and maintains exploration and discovery. It also provides an atmosphere conducive to problem solving and a place where real problems at varying levels of sophistication can be solved by pupils. Field work, for example, is an integral part of mathematics courses, from the elementary grades through graduate school.

Now let's return to the question, "What is a mathematics laboratory?" Its very nature defies a formal definition. To us, it is a concept that is based on physical materials as well as an approach to learning. As such, the mathematics laboratory approach offers a wide variety of instructional possibilities to any teacher. The mathematics laboratory is not presented as the final answer. It is recognized that the effectiveness of the instructional method or approach is largely dependent on the pedagogical capabilities of the individual teacher. However, we believe that given an improved climate for learning and assorted physical materials for student use, a teacher's effectiveness can be immeasurably enhanced.

The organizational schema of a mathematics laboratory is a function of many variables, including teacher philosophy toward learning, instructional methods, mathematics curriculum, physical facilities of classroom, as well as schools, equipment, consumable materials, money and community resources. Each of these factors is extremely important to the successful development and continuance of a situation designed to provide for the active learning of mathematics. Because of their importance, each of these factors will be discussed in subsequent chapters. It should be noted that because of the number of intervening variables, *each mathematics laboratory is and should be somewhat unique.* Hopefully one common element to these mathematics laboratories will be the approach to learning which is practiced.

POSTSCRIPT

As you might suspect, the notion of a mathematics laboratory did not develop overnight. A review of the historical development suggests a slow, irregular, and often sporadic growth. A brief discussion of the "Mathematics Laboratory in Historical Context" as well as related references are provided in Appendix E. This is optional read-

---

* Use of "mathematics laboratory" in this book assumes that the "laboratory approach" or "active learning" characterize the learning within the mathematics laboratory setting. It should be observed that our concern with the learning procedure or process provides for greater potential use of a mathematics laboratory. Every school cannot have elaborate equipment and materials for a mathematics laboratory, but every school could use this method of teaching.

ing. It will, however, give you a perspective different than the current movement toward mathematics laboratories.

SELECTED LEARNING EXERCISES

1.   Describe at least four changes in the mathematics curriculum during the last decade. Do these changes suggest that a "revolution in school mathematics" has occurred? Defend your position.

2.   Several ways to discriminate between "education" and "training" have been identified. Do you think all pupils *should* be educated? Do you think all pupils *can* be educated? Why?

3.   Should certain pupils be "trained?" If so, who? Also, if students are to be trained, what mathematical areas are of greatest importance?

4.   Does "training" or "education" as defined herein, dominate our elementary schools today? Our secondary schools? Our college classes?

5.   Examine several different definitions of a mathematics laboratory. (Check references in the chapter bibliography.) Compare and contrast several definitions. Did you find these definitions commensurate with the description of a mathematics laboratory presented in this book?

6.   Do you agree that "the only person who is educated is the one who has learned how to learn?" How can the "laboratory approach" to learning aid in producing an "educated" as opposed to a "trained" person?

7.   (Small group or class project.) Review some of the recent critical essays (such as *Crisis in the Classroom*) directed at our schools. Is their criticism directed at the curriculum or the methodology? How do you think a "laboratory approach" to learning would be viewed by these critics?

8.   (Small group or class project.) Select a mathematics topic, such as symmetry. Review materials prepared by the Minnemast Project and/or the Nuffield Mathematics Project to help teachers present this concept. From these materials, identify four specific learning activities and discuss how these activities could be implemented in a self-contained classroom.

BIBLIOGRAPHY—CHAPTER 2

Biggs, Edith. *Mathematics for Younger Children*, New York: Citation Press, 1971.

Biggs, Edith. "Trial and Experiments." *The Arithmetic Teacher*, 19: (1970): 26–32.

Biggs, Edith E. "What's Your Position on the Role of Experience in the Learning of Mathematics?" *The Arithmetic Teacher*, 18:(1971): 285–95.

Biggs, Edith E. and James R. Mac Lean. *Freedom to Learn*. Menlo Park, California: Addison-Wesley, 1969.

Bruner, Jerome S. "After Dewey What?" *Saturday Review*, June, 1961: 59–60.

Charbonneau, Manon P. *Learning to Think in a Math Lab*. Boston: National Association of Independent Schools, 1971.

Davidson, Patricia S. and Arlene Fair. "A Mathematics Laboratory—From Dream of Reality." *The Arithmetic Teacher*, 17:(1970): 105–110.

Davis, Robert B. *The Changing Curriculum: Mathematics*. Washington, D.C.: Association for Supervision and Curriculum Development, 1967.

Ewbank, William A. "The Mathematics Laboratory: What? Why? When? How?" *The Arithmetic Teacher*, 18: (1971): 559–64.

Glasser, William. *Schools Without Failure*. New York: Harper and Row, 1969.

Holt, John. *How Children Fail*. New York: Pitman Publishing Corp., 1964.

Hooten, Joseph R., Editor. "Proceedings on National Conference on Needed Research in Mathematics Education." *Journal of Research and Development in Education*, 1:(1967): 1–2.

Kidd, Kenneth P., Shirley S. Myers and David M. Cilley. *The Laboratory Approach to Mathematics*. Chicago: Science Research Associates, 1970.

Kieren, Thomas E. "Activity Learning." *Review of Educational Research*, 39:(1969): 509–522.

Kieren, Thomas E. and James H. Vance. "Laboratory Setting in Mathematics." *The Arithmetic Teacher*, 18:(1971): 585–89.

Kohl, Herbert R. *36 Children*. New York: New American Library, 1967.

Matthews, Geoffrey. "The Nuffield Mathematics Teaching Project," *The Arithmetic Teacher*, 15:(1968): 101–102.

Matthews, Geoffrey and Julia Comber. "Mathematics Laboratories." *The Arithmetic Teacher*, 18:(1971): 547–50.

Rasmussen, Lore. "Countering Cultural Deprivation Via the Elementary Mathematics Laboratory." *The Low Achiever In Mathematics*, Washington, D.C.: United States Office of Education, 1965.

Rouse, W. "The Mathematics Laboratory: Misnamed, Misjudged, Misunderstood." *School Science and Mathematics*, 72:(1972): 48–56.

Silberman, Charles E. *Crisis in the Classroom*. New York: Vintage Books, 1971.

3 THE MATHEMATICS LABORATORY IN THEORETICAL CONTEXT

INTRODUCTION

Human learning is vastly complex. It has been a concern of philosophers, scientists and educators since the time of the ancient Greeks, although the overwhelming portion of organized research and investigation within this area has been completed within the current century. The unique intellectual capability of *Homo sapiens* has resulted in tremendous modification of almost all aspects of the "natural world" and has enabled man to more fully understand and in a very real sense to control his environment. Surely this would not have been possible were it not for man's innate potential and desire to learn.

The process of learning has been a source of amazement, fascination and study for centuries. Since the mid-18th and early 19th centuries, man has systematically hypothesized, speculated about, described, investigated and researched areas germane to learning in general and to human learning in particular. Veritable mountains of material have been written about the way living things learn and the conditions under which this learning can be optimized. We shall confine subsequent discussion to various aspects of human learning.

It is perhaps ironic, given the sheer magnitude of the learning research which has been undertaken, that we still do not *know* precisely how human beings learn. Numerous theories have emerged, many describing in minute detail both the learner and the manner in which learning can be maximized. Others have chosen to adopt a comparatively broad interpretation of the learner and have satisfied themselves with a rather general description of the kinds of activities which will most effectively promote learning. Some theories have portrayed the learner as a passive recipient in the learning process, his mind as a blank slate which can be written upon at will. Others have contended that the child must be actively involved, both mentally and physically, if he is to truly benefit from a given experience. Some theorists have depicted the role of the teacher as the prime expositor of knowledge, as the person primarily responsible for children's learning. (Almost as if the child himself were not intimately

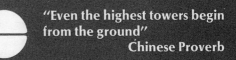

involved!) Others view the role of the teacher primarily as a guide or facilitator of learning, one who effectively organizes the conditions under which learning can take place and then exposes children to those conditions. This position tends to place more of the responsibility for learning upon the individual learner than the former viewpoint.

So we see two discernable learning theories located toward opposite ends of a continuum. This does not mean that a true dichotomy exists, as there are many theories found along the road between these two camps. In fact, most educators find themselves selecting "bits and pieces" of learning theory from each viewpoint. The way the learner and teacher perceive their roles in the actual learning process has a profound effect on the learning environment in the individual classroom. We shall examine in more detail specific implications which various theories have for the teacher of mathematics in the actual classroom. First, however, let us develop a case in behalf of the need for theories, their intended function, and lastly, describe several specific examples in the two broad camps within which most learning theories can be classified. These are the behaviorist and the cognitive theories.

THE ROLE OF THEORY IN EDUCATIONAL DECISION MAKING

Man by nature is extremely curious. He has continued, is continuing, and will continue to involve himself in activities which will help him to more fully understand the world around him. Because of his unique intellectual capability, man alone is able to wonder "why" and "how" and then proceed to systematically determine the "why" and the "how." Man's insatiable appetite for knowledge in the final analysis even transcends the practical realm of application. The incident comes to mind of George L. Mallory's response when asked why he wanted to climb Mt. Everest in the early part of this century. Sir Mallory answered, "Because it is there." Historically man has often assaulted intellectual peaks simply because "they were there." Practical benefits have often not been the prima facie reason for his intellectual assaults. The quest for new knowledge has been the sustaining force and has provided the primary source of motivation for many of man's cognitive endeavors.

Fortunately, however, results of the acquisition of new knowledge have not been confined to obscure positions on the shelves of academic libraries, but have resulted in a second major benefit for mankind; that of practical application in the real world. The development of general principles and laws has provided man with a means of predicting and controlling events in the world around him. Specific laws and/or principles tend to accumulate rapidly and their sheer magnitude often becomes cumbersome. Man very early discovered the advantages of providing a cohesive whole or generalized model from which specific laws and principles could be generated. This latter activity can be construed as the development of a scientific theory. Too often theory and practice are not considered to be related, when in fact they coexist in harmony. Hill states that "It is a serious mistake to think of a realm of theory that is separate and different from the realm of fact. It would be reasonable to say either that facts represent one kind of theory or that theories represent one kind of fact, but most reasonable to say that fact and theory represent different degrees of what is basically a single process."[1] Facts and theories, then, are inextricably intertwined, each very much dependent upon the other. Facts play a central role in the development of theory, and the theory subsequently provides a systematic interpretation of the general area to which the facts are related.

Theories have the unique capability of providing a globalized viewpoint of the segment of knowledge (or reality) under concern. They often serve as models to represent bodies of principles purposed to explain certain phenomena. Ideally, theories should provide perspective and insight to both theorist and practitioner concerned with a common area of investigation. However, one of the inherent dangers in presenting an illustrative model is to oversimplify highly complex phenomena. This oversimplification frequently distorts the real world situation and the model becomes vulnerable to attack from both theorists and practitioners.

With respect to learning theory, theorists frequently are critical of a theory because of its limited scope. That is, the proposed theory lacks universality. On the other hand, the same theories are often assailed by the classroom teacher on the basis of their "irrelevancy" to the actual classroom situation. That is, the proposed learning theory fails to provide straightforward directives to the classroom teacher.

There clearly are gaps between learning theories and classroom practice. However, what is often not realized is that consciously or subconsciously all educators subscribe to one or more ways of

---

[1] Hill, W. F., *Learning—A Survey of Psychological Interpretation*. Chandler Publishing Company, San Francisco, 1963.

perceiving the environment, the child and the teacher, and their respective roles in creating the most effective learning environment for young children. The remainder of this chapter is designed to assist educators in thoughtful mind-searching whereby their views, if subconscious, may become conscious, and if conscious may become crystallized. An integral part of this discussion will be directed toward implications which differing learning theories have for structuring activities in the mathematics classroom.

OVERVIEW

We have chosen to consider the cognitive-behaviorist distinction as an analogue of the process-product issue which is receiving much attention in some educational circles. Some segments of the science and social studies education communities have perhaps been the forerunners with respect to a serious consideration of a truly process-oriented curriculum. Perhaps in the modern world it is appropriate to depart from tradition and adopt educational goals which differ markedly from the time-honored, product-oriented objective. It might be useful to preview briefly the highlights of the forthcoming discussion where the cognitive and the behaviorist interpretation of the learning process will be considered in some detail.

To the strict behaviorist, learning takes place best in tightly controlled situations which normally preclude great amounts of student flexibility and/or variation from the "charted course." Very explicit objectives expressed in behavioral terms accompany each activity or lesson in which the student is involved. Teaching (and learning) success is dependent upon how well the student has mastered specific content material as measured by his performance, relative to predetermined evaluative criteria. Programmed instruction, although perhaps an extreme example of the behaviorist viewpoint, is indicative of the general nature of an optimal learning environment as the behaviorist perceives it. That is, the nature of student involvement in the learning process is explicitly defined with an emphasis on knowing. The criterion for success is "what" the student has learned as a result of instruction. Behaviorists for the most part assume that possession of identified prerequisite behaviors naturally implies the ability to assemble these behaviors in a manner which insures adequate synthesis, evaluative capability, and problem solving ability.

In the behaviorist tradition, transfer of what is learned is thought to be very specific. If certain skills or knowledge are required, then it is believed they should be taught directly. Involvement in activities not directly related to the attainment of an identified goal are often thought to be superfluous and therefore not encouraged. Most studies attempting to contrast the results of discovery learning and expository teaching on student achievement conclude that very tightly controlled expository teaching sequences are superior to discovery techniques when immediate learning is the goal.

Neither method appears to be significantly better when long term retention is considered.

Adherents to the cognitive viewpoint subscribe to a very different kind of environmental and pedagogical model. Although specific knowledge is not ignored, the major objectives of the cognitive position are more global and more general in nature. Whereas the behaviorist continually asks "*What* do you want the child to learn?" or "*How* do you want the child to behave?" an equally important question from the cognitive viewpoint is "*How* do you want the child to learn?" Success is as much (or more) dependent on process oriented goals as it is on the attainment of specific content (product oriented) competencies.[2] The criteria by which success is determined are, therefore, quite different. Shulman comments on some possible results of process oriented objectives:

> . . . by loosening up the objectives, we lower the probability of non-reinforcing error, while increasing the likelihood of profitable, non-threatening exploratory behavior. The error concept is irrelevant when the goal is exploration. There are goodies at every turn.[3]

Cognitive oriented learning tends to be present- rather than future-oriented. Learning activities are selected because they are of interest at the moment, and not solely because they fit neatly into the predetermined logical pattern of content development. As a result, students can find themselves immersed in problem situations for which they do not possess the necessary understandings. This assumes that children will obtain the information necessary for the problem's solution because of interest generated by the problem itself and not because it will be useful at some future time. We thus find pupils learning because of present relevance, here and now, rather than the "learn now, it will pay later" approach so often found in the mathematics classroom. The cognitive viewpoint contends that children have broad transfer capabilities, provided they are exposed to appropriate structural ideas in a meaningful learning situation. The "meaningful learning situations" usually evolve from the child's physical environment and imply active involvement (both mental and physical) on the part of the child.

Evaluation of the cognitive oriented learning sequence is often not as clear-cut and precise as in the situation where precise behavioral objectives have been predetermined. The cognitive viewpoint does not assume that the whole is equal to the sum of its parts. One must, therefore, not conclude that global objectives (i.e., structural understanding) have been attained merely because students are successful in performing tasks which indicate mastery of specific bits and pieces whose conglomerate may logically imply a coherent

---

[2] See Jerome Bruner's, *The Process of Education* (Harvard University Press, 1965) for a complete development of this point.

[3] Shulman, L., and E. Keisler, *Learning by Discovery—A Critical Appraisal*. Rand McNally, 1966, p. 97.

whole. Rather, one is encouraged to extract examples of behaviors which appear to illustrate understanding and/or knowledge of the structure of the disciplines. The lack of precision inherent in this type of evaluative process is of concern to the "cognitivist" and has been a vulnerable spot which critics of the cognitive position have frequently attacked.

Because of different criteria of success it is not possible to conduct a comparative evaluation as to whether one interpretation is "better" than the other. If one's primary intent is the immediate learning of very specific information, current research indicates that the most effective way to insure this is to teach this material directly in a tightly controlled expository sequence. If, on the other hand, the major objective is exploratory in nature, then the tightly controlled expository teaching sequence is surely inappropriate. In any well balanced mathematics program both product-oriented goals and process-oriented goals are appropriate. Teaching and learning sequences must be designed so as to promote the attainment of the objectives which have been established.

We believe the product-oriented goal and its attendant expository teaching sequences has assumed a stature exceeding its importance in the mathematics classroom. We believe it is time to involve students in a new and different type of learning environment which permits flexibility and encourages pursuit of individual interests and abilities to an extent far greater than now normally found in the mathematics classroom. We believe the mathematics laboratory to be capable of providing that kind of environment and offer the cognitive interpretation of the learning process as a theoretical rationale for its inclusion in any contemporary mathematics program.

It was suggested earlier that all educators either consciously or subconsciously subscribe to a particular conception of the child, the teacher, the learning environment, and their respective roles in the creation of the most effective learning situations. The authors will discuss essential components of the teaching-learning process from both a cognitive and a behavioral viewpoint. Although care will be taken to assure an accurate description of each of these positions, no effort will be made to conceal our own convictions. It is hoped that this chapter will help the reader recognize important advantages and limitations associated with either of the two positions. Although philosophically one may lean in one direction, it should be recognized that belief in a particular position is not an irrevocable decision. Today's teacher must be able to select appropriately from both the cognitive and the behavioral positions and adjust his conception of the teaching-learning process to each unique situation he encounters. Of course, the position which one adopts has very significant implications for the implied roles of virtually all factors related to the teaching-learning process. It is hoped that the following overview will be helpful in aiding one to clarify his point of view with respect to this very important question.

Much has been written about various interpretations of the learning process. Anything other than a brief discussion of the most salient aspects of the contemporary behaviorist and cognitive positions transcends the scope of this book. We have chosen Robert Gagne to represent a modern interpretation of the behaviorist tradition. Gagne's view of the learning process is currently receiving much acclaim within large segments of the educational community. He has had and will continue to have a non-trivial impact on the design of instruction materials at both the elementary and secondary levels. Gagne's interpretation of behaviorism is referred to as Neo-Behaviorism because of the departures he has taken from the classic behavioral position.

Jean Piaget and Zoltan Dienes represent the cognitive viewpoint. The Swiss psychologist Piaget needs little or no introduction. His theory of intellectual development is widely known, yet often misinterpreted. The overview of this theory is followed by several implications which it has for the general nature of the mathematics education of children. Zoltan Dienes has developed an international reputation in the area of mathematics learning. The basic components of his theory in this area are briefly reviewed and discussed. It is intended that the discussion to follow establish the rudiments of a theoretical justification (mathematically, pedagogically and psychologically) for the establishment of the materials laboratory in the mathematics classroom.

Readers familiar with Gagne, Piaget and Dienes or those not particularly interested in an expanded development of the Behaviorist and Cognitive positions are directed to the "Conclusions and Reactions" and the "Summary and Implications" sections following the discussions of Piaget and Dienes respectively. As the need arises in future chapters the authors will refer the reader to appropriate sections in the remainder of this Chapter.

NEO-BEHAVIORISM: ROBERT GAGNE

Ascribing to the behaviorist tradition in that his primary concern is the response of the learner following some form of instruction, Robert Gagne has been largely concerned with an attempt to clarify the relationship between the psychologies of learning and instruction. Unlike his predecessors in the behaviorist movement, Gagne is concerned exclusively with the learning patterns of *Homo sapiens.* For Gagne the central question is always: " 'What' do you want the learner to be able to do?" The degree to which the actual terminal students' behavior corresponds to the desired end result is the criterion by which school curricula are evaluated. Once the 'what' (desired terminal behavior) has been clearly defined, then it must be task analyzed. That is, one must decide which behaviors are prerequisite to the desired terminal capability, i.e., form a "learning hierarchy." It is then necessary to determine whether or not the learner possesses these necessary prerequisites. If he does, the educator may then proceed to teach the desired capability. In the event the learner does not possess all prerequisites, the needed behaviors must be conscientiously developed before the desired learning can be reached.

In a paper comparing the teaching-learning strategies of Robert Gagne and Jerome Bruner, Shulman[4] depicted the Gagne strategy diagramatically as follows:

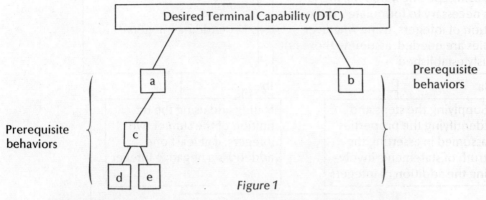

Figure 1

Gagne suggests that the prerequisite behaviors to the desired capability be assessed by answering the question, "What needs to be known before an individual could do that (terminal capability)?" When it is determined that "a" and "b" are needed—in order to be able to do DTC—the question is repeated with reference to "a" and "b". This procedure continues until all relevant prerequisite behaviors already mastered by the learner have been identified. Instruction then proceeds logically from this point, teaching the additional needed prerequisites prior to actual consideration of the desired terminal capability.

An example of such a hierarchy is shown in figure 2.

It is interesting to note that Gagne, in the true behaviorist fashion, is not concerned with "how" the desired capabilities and prerequisite behaviors are taught. Lecture, discussion, guided discovery or true discovery techniques may be utilized. The final evaluative criterion is not "how" something was learned, but rather, "what" was learned. If the learner was able to master the desired capability, the instructional strategy was successful; if he did not, the strategy used was unsuccessful. Variables other than the product or final content outcome of instruction tend to be overlooked and relegated to a position of insignificance. We thus find such factors as student motivation, positive attitudinal development, and student-teacher characteristics receiving little direct consideration. It is not so much that Gagne is totally unconcerned with these matters, but rather that they gain attention only as they relate to the promotion of the desired behavior. In all cases this criterion is identified as the product or outcome of the educative process which is usually specified in terms of *what the learner can do.*

It is not difficult to see that Gagne's model lends itself to the programming of learning sequences in a manner which divides these sequences into fragmentized or compartmentalized pieces. Implicit in this compartmentalization is the assumption that the whole is

---

4   Shulman, Lee S., "Psychological Controversies in the Teaching of Science and Mathematics". *Science Teacher*, 35:34–38, September 1968.

TASK 1   TASK 2

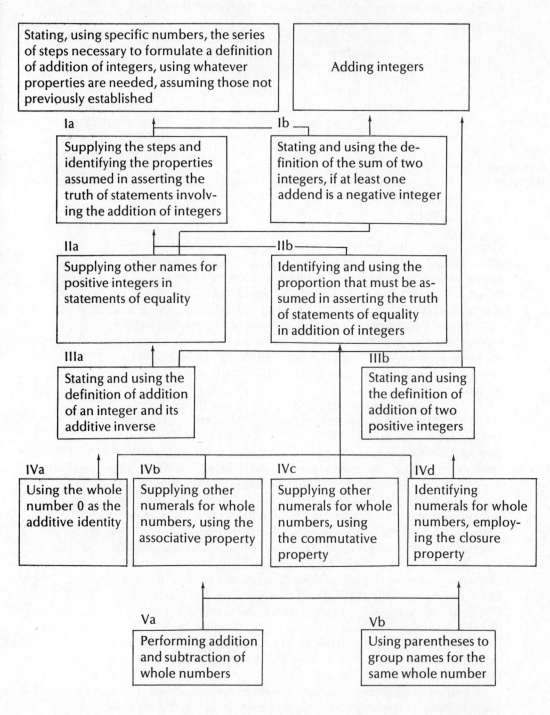

Figure 2[5]

5   Gagne, Robert M., "Learning and Proficiency in Mathematics." *Mathematics Teacher*, 56:620–26, December 1963.

equal to the sum of its parts. That is, if an individual has the prerequisite cognitive learnings he will be able to mentally assemble them in a way such that the desired terminal capability will be attained.

Gagne's insistence that educational objectives be stated in specific behavioral terms and his resulting knowledge hierarchies have formed the basis for much of the work currently being done* with the development of prime behavioral objectives for the school mathematics curriculum. When an educational program is *completely* defined in terms of desired capabilities, and the attainment of those capabilities becomes *the* major goal of the educational process, the educator may involve students in those (and perhaps only those) activities which promise to have payoff in terms of student achievement. Mathematical excursions not directly relevant to the content at hand become unacceptable. Consequently, the child's activities are in effect completely determined by the objectives (capabilities) which have been established as program goals. Such an environment often results in very limited opportunities for deviations from the development of the desired capabilities. A danger inherent in this approach is the possible lack of informal kinds of learning activities, which may not ostensively contribute to the attainment of a specific capability, but ultimately prove vital in the overall learning process. Since Gagne is primarily concerned with the "what" of the learning process, he is not particularly concerned with "how" the child learns. This does not imply however that teachers using this approach use the lecture or expository approach as the sole teaching technique. The child may be taught by lecture, or discussion, or even by discovery. The child is not *necessarily* passive (listening) and, in fact, may be quite active in the learning process. The rate at which a child progresses through the activities is not fixed. The teacher determines the pace as well as the activities and, in fact, remains accountable for the program objectives. Consequently the pupil's actions are somewhat confined, as he is likely to be involved only in those activities which will contribute most directly to the attainment of some predetermined capability or prerequisite behavior.

Using Gagne's approach to instruction, one might expect students to develop only a limited capability in the area of transfer of training. Since specific knowledges are taught directly, transfer of training is also assumed to be quite specific. Gagne believes that an individual learns what he has been taught and does not effectively apply knowledges to new situations, unless these modes of transfer

---

* See for example the materials produced by
  a) The Instructional Objectives Exchange (IOX), P.O. Box 24095, Los Angeles, California 90024 and
  b) IPI Mathematics Program (Individually Prescribed Instruction) developed through the Learning Research and Development Center at the University of Pittsburgh.
  c) IMS (Individualized Mathematics Systems) Developed through the Center for Individualized Instructional Systems in Durham, North Carolina.

are taught directly. It has been hypothesized (Shulman)[6] that this approach inhibits transfer of training because students learn specific knowledges well and these specifics in turn act as a source of interference (negative transfer) relative to the application of these specifics to new and different situations. To the degree that specific learning is done well, transfer is restricted. Although the research is inconclusive, it would appear that teaching for specific knowledges, using programmed learning, lecture method, exposition, etc., is most effective in short term specific learning situations, such as the development of computational facility.

Although Gagne shares ideas with behaviorists before him, his views are decidedly contemporary. He has had and will continue to have a significant impact on curriculum development in the area of school mathematics. His position can perhaps be best summarized by his own words ". . . there are many, many specific sets of 'readinesses to learn.' If these are present, learning is at least highly probable. If they are absent, learning is impossible. So if we wish to find out how learning takes place, we must address ourselves to these specific readinesses."[7]

THE COGNITIVE VIEWPOINT

It was suggested earlier that implicit in Gagne's model of learning is the assumption that the whole is equal to the sum of its parts. It is implied that if a person is in possession of the identified prerequisite behaviors he will be able to mentally assemble them in such a way as to assure the attainment of the desired terminal capability. Opponents of the neo-behavioristic movement take serious issue with this implied consequent. On the other hand, they would argue that in human learning, the whole is greater than the sum of its parts. In fact, they suggest that mastery of bits of knowledge does not necessarily imply the ability to organize and apply these understandings under appropriate stimulus conditions. This belief is one of the cornerstones of a second major thrust in the study of human learning. We now turn our attention to some particulars of this alternate approach.

EARLY GESTALT PSYCHOLOGY

About the time Watson published *Psychology As The Behaviorist Views It*,[8] which constituted his initial challenge to the predominant

---

6   Shulman, *loc. cit.*
7   Gagne, Robert N., "Learning and Proficiency in Mathematics," *The Mathematics Teacher*, December 1963, p. 626.
8   Watson, J. B., *Psychology as the Behaviorist Views It*. Psychological Review 1913. 20:158–177.

trends in early European and American psychology,* a second and significantly different view of the process of learning and human perception was being proposed by Max Wertheimer, a German psychologist. While Watson felt that fruitful avenues of psychological investigation should be concerned with overt behavior (in the form of stimuli and responses) rather than the more nebulous stream of human consciousness, he was in agreement with the established practice of systematic analysis of component parts or fundamental units. Wertheimer's major objection to the status quo in psychological research was not with the then current areas under investigation, but rather with the method of analysis employed. Wertheimer saw as artificial the fragmentation of consciousness into its supposed component parts for isolated analysis. He believed that consciousness should be studied as a whole so as to retain intact the characteristics which are most meaningful about it. This tendency to view reality as a whole rather than attempt the study of its component parts in relative isolation became known as Gestalt Psychology.† Gestalt Psychology is the forerunner of many current theories of learning in concert with the cognitive viewpoint.

Perhaps an analogy will further help clarify the distinction between the approaches of the behavioral psychologists such as Watson, Thorndike, Skinner, and Gagne and psychologists having a cognitive orientation. Consider a mosaic window composed of numerous individual pieces of glass of varying colors. The existing psychology in Wertheimer's time (and to a great extent, the current behaviorist point of view) views the mosaic as tiny isolated bits of color which, when taken together, comprise the scene, design or figure constituting the theme of the mosaic. This view would consider each piece of glass worthy of individual consideration prior to an attempt to view the entire window. Wertheimer (as would the contemporary cognitive learning theorist) objected to this approach. He contended that one perceives the mosaic as a meaningful whole which can be fragmentized into its component parts, but only in an artificial manner, since such fragmentation destroys the cohesion exhibited by the mosaic taken as a whole. Applying our analogy to the process of learning, the behaviorist is concerned with the components ("prerequisite knowledges" to use Gagne's terminology) of a structure and feels that only after these components are learned is one able to assemble them properly into the desired cohesive whole ("terminal capability," borrowing again from Gagne). The cognitive psychologist, on the other hand, is more concerned with

---

* These initial efforts were an attempt to define and understand the rather elusive concepts of man's sensations, thoughts, ideas and feelings. In general, early pre-Watson psychology was an attempt to study the broad parameters of human conscious experience.

† Wertheimer coined the German word "Gestalt" which can be roughly translated as "pattern," "form" or "configuration" when applied to entire systems or perceptual wholes.

an overall grasp of the entire structure and only secondarily interested in the individual consideration of the components inherent in, and attendant to, the structure itself.

It should be noted that the cognitive view of learning is often interested in the consideration of component parts; this, however, does not usually occur until *after* one has perceived the whole and is interested in how these components contribute to the overall structure of the situation. The larger picture however, is always of primary concern. Thus if one initially perceives four black dots on a sheet of paper as a square (a gestalt), this does not prevent the individual dots from retaining their separate identities. However, the primary concern of the cognitive or gestalt psychologist is the fact that these dots were originally perceived as a square. Of secondary importance is the consideration of *why* these four individual entities (dots) are perceived as a whole when possibly four other differently spaced dots do not. According to the cognitive viewpoint one cannot contend that the square is composed merely of the four dots. The positioning of the dots rather than the dots themselves is the most important factor influencing one's perception. This relationship between the entity and its components has led the gestalt or cognitive psychologists to conclude that "the whole is greater than the sum of its parts." Thus the "square" is more than merely the sum of the four individual dots.

Referring to Gagne's hierarchy (Figure 2, page 22) this interpretation would lead one to conclude that meaningful acquisition of the desired terminal capabilities (tasks 1 and 2) imply more than mere mastery of the identified prerequisite behaviors (i.e., Ia, b; IIa, b; . . . Va, b). The cognitive outlook would suggest that additional attention needs to be given to the assimilative function in the learning process.

Rather than asking "What has the individual learned to do?" the gestalt (cognitive) psychologist is more concerned with the answer to the question, "How has the individual learned to view (and/or accommodate) to the situation?" Learning is perceived here not as a process of adding new "connections" and deleting old ones, but rather as the changing or evolving of one gestalt into another. This evolution may occur as a result of actual experience or through the mental synthesis of existing structures utilizing new understandings which are often mellowed by the passage of time. The explication of the manner in which new gestalts evolve is the primary concern of gestalt learning theory.

The gestalt psychologist has a great interest in and concern for understanding. Understanding is viewed as a perception of the situation as an integrated whole, along with the manner in which means contribute to the end. This definition transcends an interpretation which equates understanding with the mere performance of logically acceptable procedures. This latter interpretation of understanding is not uncommon in educational circles. For example, consider the addition of two integers, the algorithm (rule) for long division, or the solution of simultaneous equations. Students performing these operations successfully are usually considered to "understand" precisely what is being accomplished. The authors, based on their experience in elementary and secondary schools (and also, believe it or not, at the college level) question the validity of this assumption. The ability to manipulate numerals, symbols, and sometimes even ideas is often a far cry from the degree of understanding which is ultimately desirable—no, make that minimally acceptable. At any rate, the overwhelming emphasis on understanding and on the perception of relationships as they exist within a cohesive whole has been (and continues to be) the distinctive contribution of cognitive psychology.

JEAN PIAGET

The principal goal of education is to create men who are capable of doing new things, not simply of repeating what other generations have done—men who are creative, inventive, and discoverers. The second goal of education is to form minds which can be critical, can verify, and not accept everything they are offered. The great danger today is of slogans, collective opinions, ready-made trends of thoughts. We have to be able to resist individually, to criticize, to distinguish between what is proven and what is not. So we need pupils who are active, who learn to find out by themselves, partly by their own spontaneous activity and partly through material we set up for them; who learn early to tell what is verifiable and what is simply the first idea to come to them.[9]

INTRODUCTION. Jean Piaget's contributions to the psychology of intelligence have often been compared to Freud's contributions to the psychology of human personality. Piaget has provided us with a thorough, cohesive examination and explanation of the development of human intelligence ranging from the random responses of the young infant to the highly complex mental operations inherent in adult abstract reasoning. His theory of cognitive (intellectual) development views intelligence as an evolving phenomenon occurring in identifiable stages having a constant order, although the ages of attainment of these stages may vary. According to Piaget, the age at which an individual child enters a stage as well as the time required for the child to pass through this stage is primarily a function of the following three factors: level of physiological maturation, the degree of meaningful social or educational transmission, and his relevant intellectual and psychological experiences. Piaget has developed the framework within which fruitful future research can be conducted. It should be noted, however, that there continues to exist many unanswered questions in this area of psychological research.

---

[9] *Piaget Rediscovered: A Report of the Conference on Cognitive Studies and Curriculum Development*, edited by R. E. Ripple and V. N. Rockcastle. Ithaca, New York. School of Education, Cornell University, 1964, p. 5.

The concept of equilibrium is the fourth, and in many respects, the most significant factor involved in the transition through the various stages of development. Equilibrium or auto regulation is used in a manner similar to its use in cybernetics. That is, equilibrium is reached by self-regulating processes which are cyclic in nature. "As an example, the child acquires some unorganized ideas through social learning or experience with the environment, but at some point these ideas or 'schemas' conflict, and forces must be set in motion to harmonize these conflicting ideas. This is the process of equilibration. . . ."[10] Regarding equilibrium Stendler states: "Part of man's biological inheritance is a striving for equilibrium in mental processes as well as in other physiological processes. Twin processes are involved: assimilation and accommodation. The child assimilates information from the environment which may upset existing equilibrium, and then accommodates present structures to the new so that equilibrium is restored."[11] Thus equilibrium helps coordinate the preceding three factors (maturation, experience and social transmission) and also serves as a stabilizing element in the individual's attempt to accommodate to his environment.

Early forms of intellectual development result primarily from an interaction between the child and his environment. Physical action (on the part of the child) is considered an indispensable element in the learning process. As the individual matures he becomes capable of thought processes at a more abstract level (less reliance on physical objects) until he is able to reason deductively, consider general laws and calculate the influence which multiple variables acting simultaneously have upon an argument or an outcome. In general, the "intellectually mature" individual is capable of scientific reasoning and of formal logic in verbal argument. In his book *The Psychology of Intelligence,*[12] Piaget formally develops his stages of intellectual development and the manner in which they are related to the development of cognitive structures. As a result of his concern with the nature of knowledge and the manner in which it is acquired, Piaget has been classified as a genetic epistemologist. Piaget's works transcend five decades and encompass a broad spectrum of psychological investigation. Nevertheless, we shall attempt to provide a brief summary of salient points of Piaget's research, in the hope that it will begin to lend some resemblance of structure and order to a psychological justification for the provision of a materials laboratory in the mathematics classroom.

---

[10] Copeland, Richard W., *How Children Learn Mathematics—Teaching Implications of Piaget's Research*. London: Macmillan Co. 1970, p. 19.

[11] Stendler, Celia B., "Piaget's Developmental Theory of Learning and Its Implications for Instruction in Science," edited by E. Victor and M. Lerner, *Readings in Science Education for the Elementary School. New York:* The Macmillan Co. 1967, p. 336.

[12] Piaget, Jean, *The Psychology of Intelligence.* New Jersey: Littlefield, Adams 1960

Piaget regards intelligence as the effective adaptation to one's environment. The evolution of intelligence involves the continuous organization and reorganization of one's perception of, and reaction to, the world around him. According to Piaget, the effective adaptation to one's environment (intelligence) involves the two complementary processes of assimilation and accommodation.

Piaget considers assimilation and accommodation to be biological factors since they can be found in all species. All living things exhibit a natural tendency to adapt to their environment. The specific forms of these adaptive efforts vary with the species and the constraints placed upon them by their immediate surroundings. For example, green plants grow toward a light source, since light is instrumental in regulating the direction in which the plant grows. Such growth patterns in turn maximize leaf exposure to the sun's rays which increases the process of photosynthesis. Man also exhibits a natural tendency to organize his intellectual endeavors and to adapt in some way to his immediate environment.

Assimilation is the process whereby the individual "fits" new environmental situations into his existing psychological frameworks, reacting as he has in past situations. It does not require the establishment of new cognitive structures* but merely the ability to apply existing ones to somewhat familiar situations. Assimilation can be depicted diagramatically as in Figure 3. Figure 3a represents a child's conception of the addition of two 2 digit numbers. He is confronted with a familiar situation which he has encounted before, although not under precisely the same conditions. For example, suppose the child has been adding only pairs of 2 digit numbers. He is asked to find the sum of a "new" pair of 2 digit numbers. (3b) Although he has never added these two particular numbers before he is able to follow procedures previously established and correctly carry out this exercise. We may say that he has assimilated the "new" numbers into his existing cognitive framework, which in this case is his conception of the addition of pairs of 2 digit numbers. The dotted line in figure 3c indicates his "lifespace" before he has completed the new tasks.

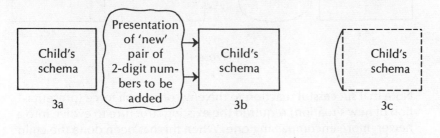

Figure 3

---

* In Piagetian language a cognitive structure is sometimes referred to as a *schema*. A schema can be thought of as an organized pattern of behavior which is consistent, orderly and coherent and is usually concerned with a specific knowing activity.

Such assimilation might be depicted by the curved solid line. Note that although the child "bulges a little at the seams", he is not forced to alter his basic structures.

When existing cognitive structures are incapable of enabling the individual to "assimilate" a new situation (that is, when existent learned responses prove to be inadequate), the individual must "accommodate" to the new situation by modifying his behavior. This modification of behavior is accomplished through a transformation or evaluation of existing structures into newer more refined and more sophisticated ones.

Consider again the example of the child who has been finding sums of pairs of 2 digit numbers. He is able to "assimilate" into his existing structures additional exercises of this type. However, when for the *first* time, he is faced with the task of finding the sum of three 2 digit numbers, he finds himself lacking the cognitive structure (schema) by which he can "automatically" carry out his assignment. He does possess schemata (plural of schema) which are relevant to the new task. He knows for example that when two 2 digit numbers are added, a new number representing the sum of the first two is produced. (Let us assume here that this sum is also a 2 digit number). Once this operation is performed, he then is faced with the familiar situation of adding two 2 digit numbers. Note that this new task requires original thought synthesis.* It requires the individual to develop a new structure based on the ones already in existence, but somewhat expanded and more sophisticated. This example of accommodation is depicted diagramatically in Figure 4.

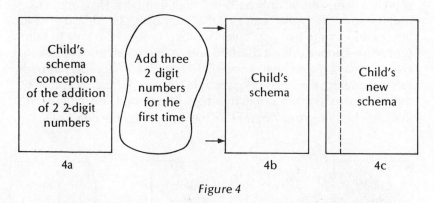

Figure 4

Note that successful reaction to the environmental force (presentation of new situation) required the existing structure to evolve into a newer, more encompassing one. When this has been done the child is able to find the sum of three 2 digit numbers by assimilating "new" situations into his newly developed schema. He is now ready to accommodate to new environmental stimuli using his newly constructed structure as a "jumping off point." We thus see assimilation

---

* A requirement for true problem solving as we use the term in this book.

and accommodation to be truly complementary in nature and constitute the manner in which, according to Piaget, human beings continually restore equilibrium within the cognitive framework. The cyclic process is depicted in Figure 5.

Assimilation

Accommodation

Cyclic Nature of Equilibration

Figure 5

Piaget contends that these processes are constantly utilized, ever renewing and continuing until the organism ceases to exist. The reader, depending upon his previous knowledge of Piaget, is either assimilating or accommodating while in the process of reading these very words. The development of intelligence is thus perceived by Piaget to be a dynamic, non-static evolution of newer more complex cognitive structures.

Piaget frequently uses the term "operation" in identifying developmental stages. Operations to Piaget are mental acts which enable one to redefine or modify his perceptions of reality. Combining, ordering, measuring, and synthesizing are a few examples. An operation never occurs in isolation. It is linked to other operations and is thereby integrated into the total mental structure.

Four main stages of intellectual development are identified by Piaget. Recall that the ages at which children progress through these stages vary, and are directly influenced by the child's maturation, experience, social transmission, and state of equilibrium. The specific ages mentioned herein are commonly accepted norms and therefore should not be construed as prescriptive for a given child.

SENSORI-MOTOR STAGE— (birth to 1½ years)—Basically pre-verbal in nature, sensori-motor intelligence is the intelligence of action. The child here is incapable of thought processes as commonly defined. During this time he is involved in the most fundamental exploration of space, time, matter and causality. These experiences develop and mold emerging mental structures. The child at this stage develops the concept of object permanence which is the notion that an object continues to exist even though it is not immediately perceivable. Toward the end of this period he can for the first time, think about objects which are not in view. This ability is the most fundamental form of symbolic representation and marks the emergence of genuine intelligence. The very young child begins to utilize low level mental combinations, such as using a chair to reach an otherwise unobtainable object from the table top. When the child moves to the pre-operational stage of cognitive development, he is capable of limited mental representation in that he can imagine some forms of the environment in a manner other than that directly observed.

PRE-OPERATIONAL STAGE—(age 1½ to 7)—Deriving its name from the fact that pre-operational children do not use logical operations in their thought processes, this stage is characterized by a process of elaboration and evolvement of additional mental operations. The pre-operational child is perceptually oriented; he makes judgments wholly on the basis of how things *look* to him rather than attempting to apply some form of reason or logic to the physical situation. The child lacks the notion of conservation of matter, and is able to keep in mind only one variable at a time, usually the one which is most prominent visually. The pre-operational child lacks the concept of reversibility, and therefore is not aware that what has been done can also be undone. That is, he lacks the notion of the relationship between an operation and its inverse. These characteristics result in the pre-operational child making logical errors. Such errors seem totally illogical and irrational from an adult standpoint, but are perfectly reasonable from the child's point of view.

The notion of conservation of matter in the pre-operational child (or rather the lack thereof) has received much attention in recent years. Piaget hypothesizes that the child has a very hazy notion of the concept of "amount." The pre-operational child cannot distinguish such variables as weight, volume, area, space, number, length, and mass. As he learns to distinguish one from the other, recognizing the unique properties of each, the child develops the cognitive structures which permit him to "conserve" each of these variables.

The following Piagetian conservation task illustrates such a confusion between the concepts of number and space:

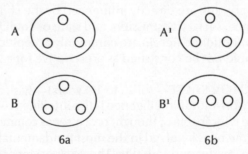

Figure 6

The child is presented with two equivalent sets of identical pieces of candy (Figure 6a) and asked to choose the set he would like to have. When asked to defend their choice most children 4 or 5 years old will reply that either set is acceptable, since they both contain the same number or "three" pieces of candy. While the child is looking, the investigator rearranges (or the child may rearrange) the candies in Set B (see Figure 6b). The same question is repeated. The pre-operational child who does not conserve number will frequently choose Set B. When asked "why" he will reply that Set B now "has more" than Set A. The child here is overwhelmed by his perceptions, he is convinced that since Set B (under rearrangement) "takes up more room" it must contain more candy. This child obviously has a very shallow conception of number, in spite of the fact that he probably is able to count to ten or more. He is confusing the concepts of number and space (or length perhaps), ultimately making his judgement on the basis of the variable which is most obvious. In this case it is the relative amount of space consumed by each of the sets.*

When applied to the pre-operational child, conservation tasks for weight, volume, length, and area will also indicate the inability to reverse operations; the inability to concentrate on more than one variable at a time, and in general, the rigidity of thought so characteristic of this stage of intellectual development. Although the rudiments of many cognitive operations emerge during the pre-operational stage, the actual structuring of these operations begins in earnest during the concrete operational stage of intellectual development.

CONCRETE OPERATIONS—(Ages 7 to 11 or 12)—At this stage the child assimilates information derived from his actions. New and improved schemata continually evolve in response to a need to accommodate new information and new situations. The child no longer reasons solely on the basis of his perceptual judgements, since thought processes are now more logical than those of the preceding stage. As a result, the type of logical error found in the pre-operational stage is no longer exhibited and the child has, for the most part, developed the concepts of reversibility and the invariance (conservation) of length, number, weight, and volume. Concrete operations receives its name from the fact that the content of the child's thinking is concrete.†

---

* Despite the fact that such reactions are commonly found in children as old as 6 and 7 (at this age a larger number of elements would probably be required) the school mathematics curriculum in the primary grades usually tacitly assumes that all children have attained conservation of number. Thus, it is not unusual to find numerous exercises of the form $1 + 1 = \square$, $2 + 1 = \square$, $5 = 2 + \square$, etc. in the first and second grade curricula. A relevant question to ask is "How can children who are unable to comprehend that the number of elements in a set is unaffected by their rearrangement, in a wholly concrete situation (candies for example), possibly "understand" this very concept embodied in a very much more abstract and symbolic form (numerals)?" Obviously they cannot. Such symbol manipulation is for many children a completely vacuous exercise devoid of all meaning. Perhaps it is time for curriculum developers to pay closer attention to the intellectual characteristics of the persons for whom they are professionally responsible.

† It should be noted that the notion of concrete is relative and exists at various levels. Thus to one child joining two beads and four beads is concrete but $2 + 4$ is not, while another may view $2 + 4$ as being concrete and $\triangle + \square$ as abstract; still later on, an older child may consider the additive group of integers to be concrete while perceiving a mathematical group to be totally abstract.

In this stage logic and reasoning abilities are utilized only as they apply to the manipulation of concrete objects. That is, he can reason logically but only about concrete phenomena. His thought processes are largely dependent upon the use of things he can see and touch,* for he has not yet developed the capability for abstract thought. Thus 8 and 9 year-olds who have no difficulty ordering a series of sticks according to length or thickness, may have great difficulty with the following type of verbal proposition. "Tom weighs more than Bill, but less than Harry; who is the heaviest of the three?" According to Piaget the ability to deal with this kind of "abstract" situation requires the ability to apply logical rules and formal reasoning. The ability to abstract problems (i.e. to think about them in an abstract context) is found only among children in the fourth stage of cognitive development, which is called Formal Operations.

FORMAL OPERATIONS—(age 11 or 12, through adulthood)—This final stage of cognitive development is initiated during early adolescence and continues to develop throughout one's lifetime. It is characterized by the emergence of cognitive structures needed for formal abstract thought. Usually by age 15 the adolescent has reached the critical stage of intellectual development in that he is capable of applying formal logic and abstract reasoning to the solution of problems. He is able to formulate hypotheses and logically deduce possible consequences from them. This ability is referred to as the hypothetico-deductive level of thought. The logical constructions of the propositional calculus (implication, conjunction, disjunction, etc.) become evident in speech patterns and are used extensively in the reasoning process. New problem situations are structured mentally before their attempted solution. This is in decided contrast to the concrete operational child who attacks problems largely on the basis of impulse and is not overly concerned with consideration of the logical implications inherent in the problem situation. Thinking at this level is less dependent upon the concrete, although this factor cannot be completely ignored. The child here can "operate with operations," that is, he is able to follow the composition of an argument while disregarding its concrete content. It is from this ability that the stage of formal operations derives its name.

The adolescent's ability to reason scientifically and to derive conclusions from formulated hypotheses is illustrated in the following experiment designed by Piaget. The individual is presented with five bottles of colorless, odorless, tasteless, and in every way, indistinguishable liquid. Three of the liquids when combined will produce a brownish solution, the fourth is a color reducing solution and the fifth a neutral element. The student is provided as much of each liquid and as many empty containers as he needs. The task is to produce the brownish solution. The "correct" solution (i.e. the one which indicates the child has made the transition to the stage of formal operations) involves the ability to systematically delineate all

---

* Certainly a strong recommendation for extensive use of audio-visual and manipulative materials in the elementary school program.

possible combinations of the liquids presented. The concrete operational child will choose bottles at random in the hope that "sooner or later" he will obtain the correct solution. The adolescent will attempt to structure this situation mentally *before* attempting to actually mix any of the liquids. He correctly reasons that if he can systematically exhaust all possible ways of combining three liquids the solution *must* eventually result. The younger child obviously has no such assurances and in fact cannot comprehend this dimension of the problem. This willingness to withhold judgement (and action) until the situation has been carefully assessed, and a method of attack formulated, is characteristic only of the formal operations stage of cognitive development.

SOME CONCLUSIONS AND REACTIONS.   In spite of the fact that Piaget's research has not been primarily concerned with the schools or with the classroom application of his findings, his genius has led to numerous insightful perspectives of the intellectual world of the individual child. These perspectives clearly have implications for educators and the manner in which they conduct the business of education. Despite the fact that Piaget has been active for more than five decades, it is only within the past ten years or so that his theories have begun to be appreciated from an educational point of view, and it will probably be at least that long before anything truly substantive (on a large scale) is done to modify the structure of the traditional classroom to truly support the intellectual development of children.*

Let us turn now to several educational implications which Piaget's work has for the school curricula in general and the mathematics curricula in particular.[13]

1.   Piaget has made it quite clear that children behave and think in a manner quite different from the adult. Children's cognitive structures, and therefore their modes of thought, language, and

---

*   Perhaps this is a bit strong, however the authors believe the traditional classroom as we know it to be incapable of providing the kind of learning opportunities that we feel are essential. It is refreshing to note that a limited number of schools in England have completely revamped their mathematics curricula (at the elementary level) to include a program largely based on student experiences. The preliminary contact which we have had with persons directly concerned with this program (The Nuffield Project) indicate an unusual degree of enthusiasm, and acceptance by all involved. Incidentally, some preliminary evidence suggests the children are learning just as much (or more) mathematics, the major difference being the way in which it is learned. Perhaps the list of implications of Piaget's research will highlight more forcefully the discrepancies between the "traditional textbook oriented program" and what we believe to be a desirable classroom setting for the mathematics program.

[13]   See also Chapter 6 of Ginsberg H. and Opper S., *Piaget's Theory of Intellectual Development*. Prentice Hall, Inc. Englewood Cliffs, New Jersey, 1969.

action differ both in quantity and quality. Each child learns differently and therefore should be exposed to a learning environment which will most effectively promote *his* individual learning style. When one observes the typical school situation, it often seems reasonable to conclude that the educational environment has been created for the convenience of the teacher rather than to maximize student learning. The point was clearly made by Furth when he remarked, "Who dares to guess how our primary education would change if teachers took seriously Piaget's proposition that knowledge is an operation that constructs its objects."[14] It has been assumed for too long that adult learning experiences can be generalized to the child. We urgently need to develop within the school environment a more refined sensitivity toward the child, his needs, and his unique learning patterns. Children are *not* little adults and therefore cannot be treated as such.

2. "Perhaps the most important single proposition that the educator can derive from Piaget's work, and their use in the classroom, is that children, especially young ones, learn best from concrete activities."[15]

The child learns through activity, both physical and mental. New structures (schemas) result from the extension of previously formed ones. According to Piaget, the effectiveness of this process is maximized when the child is actively involved in the manipulation of his environment. Of course, the physical actions or manipulations are only a means to an end, as they provide the foundation that will ultimately lead to the mental operations. If followed to its logical conclusion this finding would alter the teacher's function from one of "expositor" to one of "facilitator." That is, the teacher's role would be redefined so that his primary responsibility would have a guidance orientation. That is, he would see to the provision of a fertile learning environment and to subsequently encourage student involvement in it. Other primary functions would include the promotion of self-independence, individual exploration, and the provision of aid to individuals or small groups when needed. This conception of education is in reality quite different from the one which finds the teacher at the head of the class assuming the role of "primary fact dispenser."

3. A common misinterpretation of Piaget's four stages of intellectual development is that concrete materials are not needed by adolescent pupils. It is felt that since these individuals have passed through the concrete operational stage they no longer are in need of concrete experiences; such a conclusion is erroneous and not consistent with Piaget's findings.[16] The fact that the period of mental

---

[14] Hans G. Furth, *Piaget and Knowledge* (Englewood Cliffs, N.J.: Prentice-Hall) 1969, p. 7.

[15] Ginsberg and Opper, p. 221.

[16] Adler, Irving, "Mental Growth and the Art of Teaching" *The Arithmetic Teacher*, 13:576–584, November 1966.

growth from the ages of seven to eleven or twelve is referred to as concrete operations does not preclude the use of concrete operations at a later age. In fact, concrete operations are used at all developmental stages beyond the age of seven. Piaget's crucial point, which is sometimes forgotten or overlooked, is that until about the age of eleven or twelve concrete operations represent the highest level at which the child can effectively and consistently operate.

As adults we may be capable of operating at the formal operations level in some areas and yet find ourselves learning at a concrete operations stage when placed in a situation foreign to us. It is not uncommon to find advanced mathematics students earnestly searching for some concrete model or representation of an abstract concept. This frequently occurring phenomenon again suggests that the natural and historical development of mathematics has been largely inductive in nature, which means that it has generally proceeded from the specific to the general and from the concrete to the abstract. Likewise we find pupils who may be concrete operational in one area and formal operational at another. This phenomenon requires that teachers employ differing instructional strategies as appropriate to each child, always keeping in mind that in the development of new concepts, it is necessary to proceed from the concrete to the abstract.

4. Piaget has found that social transmission (interaction) not only affects the rate of intellectual development but can also exert a positive influence upon the quality of the child's thinking. A child assimilates his conception of reality into his existing cognitive structures in a manner which promotes a maximum degree of equilibrium. This implies that he tends to see things from his own perspective and therefore is generally not critical of his own conclusions. The opportunity to openly exchange ideas, discuss and evaluate his own ideas and the ideas of others provides him with a perspective other than his own. This leads to more critical self-evaluation and ultimately to a more objective view of reality. In Chapter 10 we will elaborate on the possibilities which the mathematics laboratory has for the promotion of this end.

5. The teacher needs to be aware of the level of intellectual functioning of each child so that appropriate experiences may be provided for him. Since children differ with respect to their rate of mental development it follows that many different learning experiences need to be available at any given time. If we are to compound this situation further by considering student interests, it is indeed possible (although not overly probable) that no two students in a given room will be involved in the same learning activity at the same time. It is difficult if not impossible to adequately provide for these needs within the traditional mathematics classroom. The mathematics laboratory is normally designed and equipped to anticipate and accommodate large differences in individual students needs and interests.

At least three time honored educational postulates (self evident truths) are questioned when a laboratory based mathematics program is seriously considered. More specifically, these "sacred cows" are:

a. All children must learn the same things,

b. All children must learn the same things at approximately the same time and in approximately the same way, and,

c. If left to their own interests students will naturally gravitate toward meaningless activity.

With regard to a. and b. above there can be no question that education is a transmission of culture and that there exists certain aspects of it with which all students should be familiar, but is it not ludicrous to expect all students to learn the same things in basically the same manner? We believe that students as unique individuals need to be given many opportunities to make decisions, which might very well begin by providing them with some degree of autonomy in the structuring of their own education. When a student is immersed in a truly fertile environment and given the opportunity to select the manner in which he will ultimately interact with that environment, then his autonomy becomes evident while at the same time educational "control" is exerted.

This is tantamount to describing the educational parameters within which the student will function and providing him with freedom of choice within those parameters. The true challenge to educators is the development of that kind of environment. If it can be constructed in such a way that whatever activity is selected results in a desirable avenue of intellectual pursuit, then we will in effect have eliminated a major stumbling block retarding educational progress.

The last axiom stated above c., although widely accepted (we know it is widely accepted because for the most part every single minute of the child's school day is pre-planned for him) raises serious questions as to the degree of faith the educational community has in students' capacity for self-direction, internal motivation to learn, and intellectual capabilities. All children are born with the desire to learn. The real challenge for educators is to find ways to facilitate and motivate that innate desire.

Children learn most effectively when they are actively involved, and that which is to be learned has meaning and interest for them. What better way to assure that these criteria are met than to encourage students to assume some degree of responsibility for their own education? There will be problems initially, to be sure. However we believe that the short period of apparent chaos will pay large dividends as students increasingly develop the capacity for effective self-direction. Mathematics with its wide diversity of ideas and concepts is especially well suited to the promotion of these ends. The mathematics laboratory can provide the environment within which these goals can be realistically attained.

In conclusion, we feel that within Piaget's theory of Cognitive Development lie the basic components of a theoretical justification for the provision of a mathematics laboratory. Zoltan P. Dienes, while generally espousing the views of Piaget, has made contributions to the cognitive view of mathematics learning which are distinctly his own. The work of this mathematician/psychologist/educator lends further support to the concept of a laboratory-based mathematics program.

ZOLTAN P. DIENES

INTRODUCTION. Dienes* views the effective learning of mathematics to be dependent upon the establishment of self-motivating learning situations. These situations evolve from the child's experiental world and should be designed to explicate the structural aspects of the subject in game-like situations which have both meaning and relevance for him. Although this laudable goal would be acclaimed by most mathematics educators (regardless of their philosophical and psychological orientation), the means of achieving this places him squarely in the cognitive camp. His perceptions are largely commensurate with other contributions to the cognitive view of mathematics learning. Dienes' strong background in both mathematics and psychology has provided him with the capability to develop a theory of mathematical learning which is sound from both a mathematical and psychological point of view.

Rejecting the position that mathematics is to be learned primarily for utilitarian or materialistic reasons (i.e. to increase the probability of attaining a higher level economic position of life), Dienes sees mathematics as an art form to be studied for the intrinsic value of the subject itself.[17] The learning of mathematics should ultimately be integrated into one's personality and thereby become a means of genuine personal fulfillment. Noting the serious, and perhaps even desperate, status of the past and present situation with respect to the attainment of mathematical understanding by the population at large, he suggests a serious re-examination of the relevancy and effectiveness of established practices in the mathematics classroom. Making a clear distinction between the acquisition of techniques and the understanding of ideas, he sees current practices primarily oriented toward the promotion of the former and only superficially concerned with the development of concepts.

---

* Of Hungarian descent, Zoltan Dienes spent his early professional life in England and Australia. After receiving his Ph.D. in Mathematics he became interested in the study of logic and the psychology of cognition. His educational interests have led to the publication of several books and articles in which he addresses himself to the problems of mathematics learning. The rudiments of his proposed theory will be discussed in this section. Professor Dienes is currently the Director of the Institute for the Study of Mathematics Learning at Sherbrooke University in Quebec, Canada.

[17] Dienes, Z. P. *Building Up Mathematics*. Hutchinson Educational Ltd., London, England, Revised Edition, 1969.

Dienes views the perceived necessity of having an externally imposed punishment-reward (grading) system as clear evidence of the failure of schools to develop the internal motivational systems believed to be contained in all children. The artificial nature of having students pursue high grades in order to avoid punishment (poor grades) is clearly less than an optimal way to encourage and promote a genuine desire to learn and to become truly involved in an educational experience. Little comfort can be derived from the fact that this situation is not confined solely to the area of mathematics.

The present conception of education as a process whereby information is disseminated to large groups of students at the same time and in a similar manner is almost totally inadequate and in vital need of re-organization. The traditional instructional model does not recognize the fact that learning is primarily an individual matter and as such is perhaps not amenable to the class-lesson paradigm. Dienes is in a sense rejecting the very pillars upon which the present educational system is based; from its objectives to its methodology, from its subject matter to the instructional materials utilized, and from the qualifications of its personnel to the constraints placed upon the physical structure housing the learning environment.

These larger and more general concerns raised by Dienes are indeed worthy of detailed consideration by all members of the educational community; however for our purposes here, let us now turn our attention to his theory of mathematics learning, noting its implications for the restructuring of the mathematics learning environment.

Dienes views the learning of mathematics as a process of evolution whereby existing understandings form the basis upon which new and more complex ones are developed. The learner continually attempts to relate newly acquired structures to those already within his grasp in a manner which absorbs and processes the new information. These procedures ultimately redefine his "mathematical life space" to permit the processing of new environmental stimuli.*

Dienes defines mathematics as "actual structural relationships between concepts connected with numbers (pure mathematics), together with their applications to problems arising in the real world (applied mathematics). He views mathematics learning as "the apprehension of such relationships together with their symbolization, and the acquisition of the ability to apply the resulting concepts to real situations occurring in the world". Dienes ques-

---

* This is an example of Piaget's complementary processes of assimilation and accommodation directly translated into the area of mathematics learning. It is further evidence of the impact which Piaget has had on the cognitive viewpoint. The reader will note additional applications of Piaget's work in the further discussion of Dienes' position.

tions the appropriateness of the stimulus-response model as a viable means of promoting mathematics learning. This rejection is predicated on the contention that mathematics is concerned with the structure of relationships and as such should be approached in a manner which facilitates the recognition and understanding of its "big ideas." The S-R (Stimulus-Response) model is not felt to be particularly conducive to this kind of global involvement, since it is by nature concerned primarily with the mastery of specific content and therefore can more easily degenerate into a "bits and pieces" approach.

Dienes' theory of mathematics learning has four major components or principles. We shall discuss them separately noting some implications of each.

THE DYNAMIC PRINCIPLE. The dynamic principle suggests that true understanding of a new concept is an evolutionary process involving the learner in three temporally ordered stages.

The preliminary, or play stage, is in evidence when the learner is concerned with activities of a relatively unstructured nature but which provide actual student experience to which later experiences can be related. Although unstructured, the activities here are not random. The child is presented with an "educational ball park," within which the concept(s) are embedded, and given complete freedom to "play" in it. The learner is therefore exposed to basic rudiments of concepts in a very informal manner. Since freedom to experiment is essential, the learners' responses, varied though they may be, must be relatively uninhibited by the teacher and treated with genuine respect. One sees this stage in action when kindergarten or first grade children are presented with Cuisenaire rods for the first time. The natural tendency of young children (and even pre- and inservice elementary teachers) is to construct trains, build castles or formulate designs with the rods. (These designs, by the way, often turn out to be symmetrical and usually suggest a natural ordering system for the rods.) This is evidence of the innate human tendency to lend some semblance of order to a seemingly chaotic situation.

Following the informal exposure afforded by the play stage, more structured activities are appropriate. It is here that the child is given experiences structurally similar to the concepts to be learned. To use the Cuisenaire rods as an example, the child might be asked to make a train just as long as a black rod in as many ways as he can. The possibilities are illustrated in Figure 7. It is obvious that the concept ultimately to be considered will be addition, and that this activity is laying the foundation for the future notion that a number can be expressed in many different ways; we can write $7 = 1 + 6$ or $7 = 2 + 5$ or $7 = 3 + 4$, etc. It is significant to note that the child initially is not (or should not) be concerned with a 1 bar or a 6 bar or with a 2 bar and a 5 bar, but rather, with a white bar and a dark green bar or with a red bar, and a yellow bar, respectively. He is thereby experiencing the concept of addition in a considerably simpler and more concrete form (joining rods), a form which will later become the foundation upon which his abstract conceptions of addition (i.e., $2 + 5 = 7$) are based.

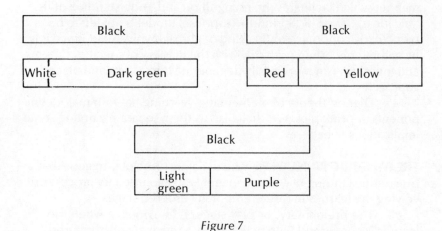

Figure 7

WE DIGRESS: This notion of similarity between structural systems (in the above example, the rods form one system and the numbers the other) is extremely important from both a mathematical and a pedagogical point of view. In mathematical terms this structural similarity between two systems is known as an isomorphism.* Mathematically, the concept of isomorphism affords a great deal of economy of effort, for one can study the properties of one system, identify another system isomorphic to it, and immediately transfer all one has learned about the first system to the second. Pedagogically the notion of isomorphism enables us to approach relatively abstract ideas in a concrete manner, because we know the two systems "behave" in precisely the same manner and that an operation within one system will have its counterpart in the other. Using the previous example, the property of length as contained in the Cuisenaire rods is isomorphic to a subset (positive integers i.e. 1, 2, 3, 4 . . .) of the real number system. For every length rod there is a corresponding unique number so that the operation of "joining

---

* Technically, the two systems of $a, b, c, \ldots$ and $\alpha, \beta, \gamma \ldots$ are isomorphic if an equality and an operation are defined and the sets are in one to one correspondence. Thus if $a \Leftrightarrow \alpha$ (read a is equivalent to alpha), $b \Leftrightarrow \beta$ (beta), $c \Leftrightarrow \gamma$ (gamma) . . . such that:
1) $a = b$ if and only if $\alpha = \beta$
2) $a + b \Leftrightarrow \alpha + \beta$
Then we can say that the two systems are isomorphic (structurally identical) relative to the operation of addition.
If in addition to 1) and 2) above we impose the third constraint 3) $ab \Leftrightarrow \alpha\beta$ then the two systems $a, b, c \ldots$ and $\alpha, \beta, \gamma \ldots$ are said to be isomorphic with respect to both addition and multiplication.

rods" in the concrete system corresponds to that of "adding" numbers in the more abstract one. This is illustrated in Figure 8.

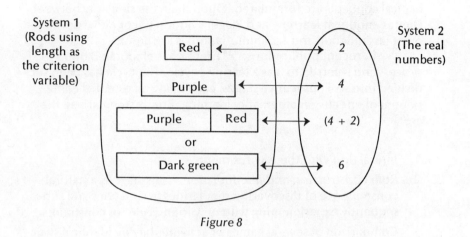

System 1
(Rods using length as the criterion variable)

System 2
(The real numbers)

Figure 8

Extending the above example, it has been demonstrated that children are able to grasp relatively sophisticated ideas (associativity, inverse operations, distributivity, for example) about a completely abstract system (real numbers) at an early age, because the ideas were initially embodied in a simple, more concrete system which is isomorphic to the abstract one. The exploitation of isomorphic systems (analogous to finding appropriate models) for purposes of instruction is not germane solely to the primary grades or even to the elementary school. It is in fact a basis for conceptual learning at all levels of instruction.

RETURNING TO DIENES. The third component of the dynamic principle, following the more structured "becoming aware" stage, is characterized by the emergence of the mathematical concept with ample provision for reapplication to appropriate environmental stimuli. The concept is not considered to be fully operational until it can be freely recognized and applied to relevant situations. Ideally this practice stage will serve a dual role: to solidify the newly formed concept in the child's experience and to serve as a play stage for the next concept to be learned. A cyclical pattern emerges which can be depicted as follows:

Play

Abstractions reapplied
to the environment from
whence they came, forming
play stage for new concepts.

Structuring of
informal play
activities

Abstractions

The completion of this cycle is necessary before any mathematical concept becomes operational for the learner. Furthermore, Dienes maintains that this cycle is continuously repeated as new mathematical concepts are formulated. Once again the similarity between Dienes' model of learning and Piaget's notion of cognitive development (assimilation and accommodation) is apparent.

In recent publications,[18],[19] Dienes has elaborated upon this process and referred to it as a learning cycle. The cycle has been subdivided into six consecutive stages, each being an essential component of effective mathematics learning. Briefly summarized they are:

1. *Interaction* with the environment
2. *Rule construction and manipulation*—this stage is a natural consequence of discovering regularities in situations and subsequently experimenting with new found rules or constraints.
3. Comparison of several games or activities having identical structures or rules. This is referred to as the search for *isomorphisms*.
4. *Representation* of isomorphic structures—a realization of their "sameness," together with an ability to determine a method of representing all games (activities) having a similar structure.
5. *Symbolization*—The investigation of the properties identified in #4. Such investigation at this level is not dependent upon the real world referent, although it should be available to corroborate findings generated in symbolic form.
6. *Formalization*—the derivation of other properties of the system from those already identified. This is the process of generating theorems (statements logically deduced) of a system from its axioms (rules of the game or self-evident truths). The process of theorem derivation is known as a proof and is considered to be the quintessence of mathematical activity.

The reader interested in a more complete description of these stages together with specific examples should consult the references indicated.

These stages describe the general nature of a sequence of experiences which result in the appropriate development and subsequent abstraction of a given concept. The remaining components of Dienes' theory are in a sense contained within the dynamic principle, and should be considered to exist within the framework already established.

---

[18] Dienes Z. P., Golding E. W., *Approach to Modern Mathematics*. Herder and Herder Co. New York, 1971.

[19] Dienes, Z. P. "An Example of the Passage from the Concrete to the Manipulation of Formal Systems," from *Educational Studies in Mathematics*. D. Reidel Publishing Co. Dordrecht, Holland, 1971, pp. 337–352.

THE PERCEPTUAL VARIABILITY PRINCIPLE*. In an attempt to make provision for individualized learning styles, Dienes suggests the perceptual variability principle as an indispensable element in the process of concept formation. This principle suggests that conceptual learning is maximized when children are exposed to a concept in a variety of situations. The experiences provided should differ in outward appearance while retaining the basic conceptual structure. Piaget and others have shown that children view the world differently than an adult and many times will become sidetracked with irrelevant characteristics of a situation.

> As adults, with some knowledge, our approach to many things is abstract. We readily ignore some characteristics in order to be satisfied that the remaining properties will do something for us. . . . Children do not ignore the different characteristics of materials so readily. In fact, we could say that the ignoring of different characteristics is directly proportional to the degree of knowledge one already has or is expected to have. This means that for any given situation which may seem to be a reasonable starting point for a discussion, the attention the teacher pays it and the attention the children pay it will be different and this can give rise to difficulties in comprehension—each misunderstanding the other.[20]

Dienes believes that the current mode of mathematics instruction promotes learning which is associative in nature. Children are encouraged to associate a particular mathematical process or operation with a particular situation.† A "bag of tricks" approach is often used by the teacher, applying established mathematical procedures in a manner which tends to be routine or habitual in nature. The problem with associative learning arises when the student finds himself in a situation for which he does not possess a ready-made response pattern. Under these conditions the associative learner often finds himself bewildered and unable to abstract relevant components of the problem situation. It is, therefore, unlikely that he will be able to reconstruct the problem situation in a manner which will ultimately lead to a solution. As an alternative to this type of learning, Dienes suggests an instructional setting which would promote abstraction rather than association. (Abstraction is here defined as the ability to perceive a concept irrespective of its concrete embodiment.) Dienes believes that by providing children with the opportunity to see a concept in different ways and under different

---

* Also known as the "Multiple Embodiment Principle." Although the case for "multiple embodiment" is pedagogically sound and has been ably presented by Dienes, it has yet to receive widespread use by classroom teachers.

[20] Association of Teachers of Mathematics, *Notes on Mathematics in Primary Schools*, Cambridge University Press. London: 1967, p. 168.

† Very similar to a stimulus-response connection but could be more global in nature such as B. F. Skinner's operant conditioning.

conditions, the purposes of promoting abstraction will be more adequately served.

As an illustration, the concept of place value implies a numeration system in which the value of a symbol is dependent upon its position relative to a reference point. (In the decimal system, base ten, this reference point is called a decimal point.) Consider the concepts of addition and subtraction which obviously can be depicted in many different ways, some of which follow. These examples relate only to base ten although they could be revised to accommodate other bases.

a.   with small sticks and rubber bands in assorted piles

b.   with poker chips assigning values to various colors (i.e., 10 white chips = 1 red chip etc.)

c.   with an abacus (10 beads to a string)

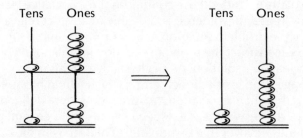

d.   with a place value chart using small cardboard cards and rubber bands

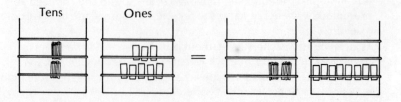

e. with the Multibase Arithmetic Blocks (MAB)*

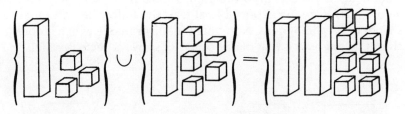

f. with numerals using the traditional algorithm, i.e.,

$$\begin{array}{r} 13 \\ + 15 \\ \hline 28 \end{array}$$

Dienes views a. through f. above as different embodiments of a single concept and suggests that the student be exposed to as many as are required for the student to develop the ability to abstract. The number of embodiments needed will vary from child to child. It is felt that when children are exposed to a number of seemingly different tasks which are essentially identical in structure, they will tend to abstract the similar elements from their experiences. It is not the performance of any one of the individual tasks which is the mathematical abstraction but the ultimate realization of their similarity. Provision for perceptual variability is, therefore, felt to be an essential component of a viable instructional strategy.

THE MATHEMATICAL VARIABILITY PRINCIPLE. Similar to the Perceptual Variability Principle in that it encourages multiplicity in patterns of exposure, the mathematical variability principle asserts that if a mathematical concept is dependent upon a certain number of variables (quantities which may assume a variety of values), then variation of these is an important prerequisite for the effective learning of the concept. For example, a parallelogram is defined as a quadrilateral having its opposite sides parallel. Although its shape can be varied by changing the length of its sides and the size of its angles, the only crucial factor is that its opposite sides remain parallel. If a child were exposed to only those parallelograms with equal angular measure or only to those which have constant proportion relative to the length of its adjacent sides, surely he would not develop a general concept of parallelogram. The mathematical variability principle suggests that in order to maximize the generalizability of a mathematical concept, as many irrelevant mathematical variables as possible (in this example the size of angle, length of side and position on paper) should be varied while at the same time

---

*   These blocks have been designed by Dienes to provide one embodiment of the concept of place value. The MAB materials provide for bases 2, 3, 4, 5, 6 and 10. All are used in conjunction with the rules of place value and would be utilized in a manner similar to that employed with materials listed in a., b., c., and d. above. Chapter 7 provides laboratory activities utilizing the MAB materials.

keeping the relevant variables (opposite sides parallel) intact. A student in the process of formulating a concept of parallelogram would, therefore, be exposed to at least the following different shapes of parallelograms. Note that the only variable which is consistent throughout is that of opposite sides always being parallel.

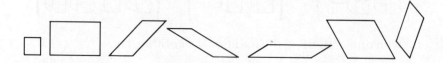

By the same token, if one is interested in promoting an understanding of place value, it is desirable to vary the base while providing experiences which highlight the consistency of regrouping procedures, the importance of relative positional value, and the appropriate way to record results. When implemented, the mathematical variability principle encourages the student to separate the "wheat from the chaff" by systematically isolating relevant variables in the consideration of a given concept. This results in conceptual learning with a degree of precision transcending that usually found in the mathematics classroom.

THE CONSTRUCTIVITY PRINCIPLE.   Here Dienes identifies two kinds of thinkers: the constructive thinker and the analytical thinker. He roughly equates the constructive thinker with Piaget's concrete operational stage and the analytical thinker with Piaget's formal operational stage of cognitive development. This principle states simply that "construction should always precede analysis." It is analogous to the assertion that children should be allowed to develop their concepts in a global intuitive manner emanating from their own experiences. According to Dienes these carefully selected experiences form the cornerstone upon which all mathematics learning is based. At some future point in time, attention will be directed toward the analysis of what has been constructed; however, Dienes points out that it is not possible to analyze what is not there.

SUMMARY AND IMPLICATIONS.   The unifying theme of these four principles is undoubtedly that of stressing the importance of learning mathematics by means of direct interaction with the environment. Dienes is continually inferring that mathematics learning is not a spectator sport and, as such, requires a very active type of physical and mental involvement on the part of the learner. In addition to stressing the environmental role in effective conceptual learning, Dienes in his two variability principles addresses the problem of providing for individualized learning rates and learning styles. His constructivity principle aligns itself closely with the work of Piaget and suggests a developmental approach to the learning of mathematics which is temporally ordered to coincide with the various stages of intellectual development.

If the work of Zoltan Dienes were to be taken seriously, a very different kind of classroom model would be employed. The "class-lesson" would be virtually eliminated in order to accommodate individual differences in ability and interests. Almost all activities would be designed for individuals or for groups of two or three, since it is not likely that more than this number of children will be ready for the same experiences at the same point in time. Such an organizational scheme obviously has vast implications for the role of the teacher, since it is no longer possible for him to be the primary source of information for all groups simultaneously. The student's role in the learning process is also dramatically changed for he now must assume a greater degree of responsibility for his own education. This situation creates new demands for additional sources of information and direction. Dienes suggests the creation of a learning laboratory which would have, in addition to a large assortment of physical material, a system of assignment cards arranged both in "series"—development of a concept using a number of related tasks —and in "parallel"—presenting a given concept within a variety of embodiments. These cards should offer both choice and variety to accommodate differences in student abilities and interests. In short, a radical overhaul of existing pedagogical practices, teacher-pupil interaction patterns, as well as the general aspects of classroom climate, is required by this view of mathematics learning.

The reader will notice a large degree of similarity between what is implied here and that which was inferred from the work of Jean Piaget. As was suggested earlier, Dienes, in developing a theory of mathematics learning, has leaned heavily on the work of Jean Piaget. We believe that Zoltan P. Dienes has made an important contribution to the understanding of mathematics learning. The mathematics learning theory proposed by Dienes, although undoubtedly incomplete, can provide a basis from which other researchers and theoreticians can launch subsequent investigations. In addition to its theoretical implications, there exists a truly practical side to Dienes' ideas. For in a very real sense his work provides a vehicle whereby the theoretical can be interpreted and implemented in the practical classroom situation. Chapters 6 and 7 provide exemplary series/parallel laboratory lessons designed to illustrate Dienes' theory of mathematics learning.

We turn now to the "real world" of the mathematics classroom in an attempt to provide an indication of the manner in which a mathematics laboratory can actually be implemented. The discussion and suggestions made reflect our experience with the laboratory approach and have proved effective for many classroom teachers at both the secondary and elementary levels.

SELECTED LEARNING EXERCISES

1.   Robert Gagne has proposed a procedure which can be used to determine the logical content prerequisites necessary to the understanding of higher order concepts. The procedure is often referred

to as task analysis. An example was discussed earlier in this chapter. Develop a task analysis similar in format to the one mentioned above for one or more of the following Desired Terminal Capabilities.

a. The ability to classify polygons according to the number of sides. (i.e., triangles, quadrilaterals, pentagon, etc.)

b. The ability to compute the area of a triangle using the formula $A = \frac{1}{2} bh$.

c. i) The ability to find the sum of 2 one digit numbers whose total does not exceed 9.

    ii) The ability to find the sum of any pair of one digit numbers.

    iii) The ability to find the sum of 2 two digit numbers whose sum does not exceed 99.

    iv) The ability to find the sum of any pair of two digit numbers.

2. Piagetian conservation tasks have proven useful in determining the intellectual level at which children are operative. They are easy to administer and generally require materials which are readily available. All of these tasks should be done in an interview setting with a single child. Children normally conserve number, length and volume between the ages of 6 and 8, although exceptions are not uncommon. Children should have the opportunity to "handle" the materials and should always be asked to justify or explain their conclusions.

You are asked to try each of these experiments with several different children keeping a careful record of their ages and responses in order to contrast your results. Total testing time for each child should not exceed 15 minutes.

a. CONSERVATION OF NUMBER: I
Appropriate age of child—5 or 6.

MATERIALS NEEDED—10–12 identical counters.

PROCEDURE: Arrange counters in two identical sets. i.e.,

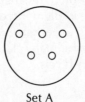 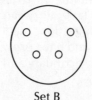

Set A   Set B

Ask child whether both sets contain the same number of counters or if one set has more. (It is important at this point that he agree that both sets contain the same number of elements). While child is observing, re-arrange Set B as follows;

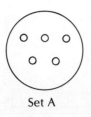 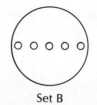

Set A   Set B

Ask child if there are still as many elements in Set B (after re-arrangement) as there are in Set A. Remember to have child explain his answer. If the child is a pre-conserver with respect to number his answers will surprise you.

b. CONSERVATION OF NUMBER: II
Appropriate age of child—6 to 8

MATERIALS NEEDED—20–40 identical counters, 2 containers.

PROCEDURE: Two empty containers are placed before child. The containers are "filled" with the counters one at a time. Be sure that each time a counter is placed in container A, a counter is *simultaneously* placed in container B. After each 5 or 6 pairs of counters are so placed, ask the child whether or not each container holds an equal number of counters. It is important that he agree that each contains the same number but it is not necessary to count them. When each container holds 20 or so counters, pour the contents of one of them onto the table or desk top, spreading them over a large area. Then ask, "Are there just as many counters on the table top as there are in the other container, or are there more counters on the table top or are there less counters on the table top?" Ask the child to explain.

c. CONSERVATION OF LENGTH:
Appropriate age of child—5 to 8

MATERIALS NEEDED—Two pencils, rods or strips of paper having identical lengths.

PROCEDURE: Place two objects side by side as in diagram below.

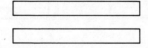

Ask child if they are of equal lengths or if in fact one is longer. Remember it is important that he agree they are equal in length at this point. If he does not, find two things which he will agree are equal in length. New simply translate one of the objects to the right or have him do so, i.e.,

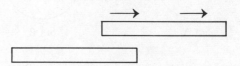

Repeat the question, "Are they still the same length?" Ask him to justify his answer.
Variations on the same theme include:

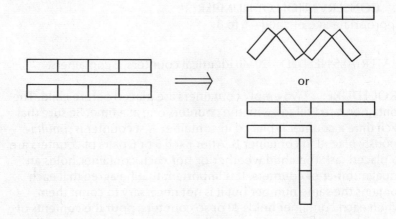

or

d. CONSERVATION OF VOLUME:
Appropriate ages—5 to 8

MATERIALS NEEDED: Water, two identical containers. 1 container with lesser diameter than other two.

PROCEDURE: Fill two identical containers to the same level, i.e.,

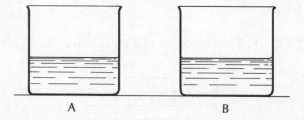

Ask, "Do both contain the same amount of water?" (You may need an eye dropper to make the necessary adjustments so that he answers yes to this question.)

Now pour container B into the third container, i.e.,

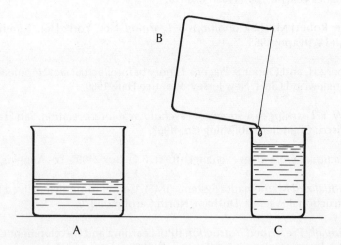

Ask "Is there just as much water in container C as there is in container A? Why or why not?"

After you have completed these Piagetian tasks with several children and have recorded your findings, find a classmate who has done the same. Compare your results. What implications do these results have for you, the classroom teacher, in so far as the teaching of mathematics (and science) is concerned?

BIBLIOGRAPHY—CHAPTER 3

Adler, Irving. "Mental Growth and the Art of Teaching." *The Arithmetic Teacher*, 13:(1966): 576–584.

Association of Teachers of Mathematics. *Notes on Mathematics in Primary Schools*. Cambridge University Press. London: 1967.

Bruner, Jerome. *The Process of Education*. Harvard University Press, 1965.

Copeland, Richard W. "*How Children Learn Mathematics—Teaching Implications of Piaget's Research*." London: Macmillan, 1970.

Dienes, Z. P. *Building Up Mathematics*. (Revised Edition) London. Hutchinson Educational, 1969.

Dienes, Z. P. "An Example of the Passage from the Concrete to the Manipulation of Formal Systems." *Educational Studies in Mathematics*. 3: (1971): 337–352.

Dienes, Z. P. and E. W. Golding. *Approach to Modern Mathematics*. New York: Herder and Herder Company, 1971.

Furth, Hans G. *Piaget and Knowledge*. Englewood Cliffs, N.J.: Prentice Hall, 1969.

Gagne, Robert M. "Learning and Proficiency in Mathematics." *Mathematics Teacher*, 56:(1963): 620–26.

Gagne, Robert M. *The Conditions of Learning*. New York: Holt, Rinehart and Winston, 1965.

Ginsberg H. and Opper S. *Piaget's Theory of Intellectual Development*. Englewood Cliffs, New Jersey: Prentice Hall, 1969.

Hill, W. F. *Learning—A Survey of Psychological Interpretation*. San Francisco: Chandler Publishing Co., 1963.

Instructional Objectives Exchange (IOX), P.O. Box 24095, Los Angeles, 1970.

Individualized Mathematics Systems (IMS). Center for Individualized Instructional Systems. Durham, North Carolina, 1970.

Individually Prescribed Instruction (IPI). Learning and Development Center. Pittsburgh: University of Pittsburgh, 1968.

Piaget, Jean. *The Psychology of Intelligence*. Littlefield, New Jersey: Adams, 1960.

Piaget Rediscovered: A Report of the Conference on Cognitive Studies and Curriculum Development, edited by R. E. Ripple and V. N. Rockcastle. Ithaca, New York: School of Education, Cornell University, 1964.

Reys, Robert E. "Mathematics, Multiple Embodiment and Elementary Teachers" *The Arithmetic Teacher* 19: 400.

Shulman, Lee S. and Keisler Evan, R. *Learning by Discovery—A Critical Appraisal*. Chicago: Rand McNally, 1966.

Shulman, Lee S. "Psychological Controversies in the Teaching of Science and Mathematics." *Science Teacher*, 35:(1968): 34–38.

Stendler, Celia B. "Piaget's Developmental Theory of Learning and Its Implications for Instruction in Science." edited by E. Victor and M. Lerner. *Readings in Science Education for the Elementary School*. New York: The Macmillan Co., 1967.

4 THE MATERIALS LABORATORY—
SOME PRACTICAL CONSIDERATIONS

OVERVIEW

This chapter will address the problem of organizing physical resources within the mathematics classroom in order to create an optimal climate for effective student learning experiences. The classrooms described are based on a child-centered approach to the learning of mathematics, necessary for the effective implementation of the materials laboratory in the mathematics classroom.

If one is to attempt to develop a laboratory based mathematics program with emphasis on individualized learning, a different kind of classroom model is needed. For a start, an informal atmosphere should prevail. Pupils should be given the freedom to move about as they seek answers to *their* questions and should be continually urged to assume responsibility for their own learning.* The teacher must, therefore, adopt the role of "facilitator" and at least partially abandon the traditional role of "expositor." The physical arrangement of classroom furnishings must be altered to promote pupil freedom of movement and differentiated mathematical activities. Additional instructional materials will most probably have to be procured to provide the needed instructional support base for the newly designed program.

At this point, many questions naturally arise. For instance, how does one initiate an informal atmosphere in the mathematics classroom? How can the responsibility for learning be shifted from teacher to student? How should classroom furniture be rearranged to facilitate the laboratory approach? What kind of materials are needed? How does a teacher begin to implement an activity oriented method of learning? What are some practical suggestions to help assure suc-

---

* This is not to be confused with total freedom where students in effect make all decisions as to what will be learned and how it is to be learned. Student freedom when carried to this extent is felt to be an abdication of professional responsibility. We envision instead freedom within parameters which have been established by the teacher. (Hopefully in collaboration with students.)

To teach is to learn twice.

cess during the crucial early stages? It is toward these and other important questions that the following discussion is directed.

The organization of the human component in the mathematics laboratory is considered in Chapter 9.

CLASSROOM DESIGN

INTRODUCTION. Lack of adequate classroom facilities frequently restricts the development of activity oriented lessons. Ostensibly it is this lack of necessary physical materials and equipment that precludes the development of laboratory activities. However, upon more careful examination, one usually finds that effective laboratory activities can be implemented and effectively utilized by teachers with some degree of ingenuity, creativeness, and enthusiasm for a laboratory approach to learning mathematics.

Even under relatively favorable conditions, the difficulty encountered in conducting laboratory activities in a traditional classroom setting can cause real problems. These problems are further compounded by the frequent lack of adequate supportive physical facilities. Although there is no apparent one best solution in so far as mathematics laboratory facilities are concerned, there are certain classroom designs and physical facilities which are felt to be best suited to a laboratory approach. It is, of course, recognized that teachers will need to adapt these suggestions (or modifications of them) to their own schools.

Since the classroom is the focal center for most learning activities it must provide an atmosphere that will foster "bona fide" learning activities. The classroom must provide such essentials as adequate light, proper acoustics, comfortable temperature, proper ventilation, and adequate space. In addition, the mathematics classroom should have an ample amount of quality chalk and bulletin boards, screens for both overhead and movie projectors, and furniture. Although the mere mention of such classroom facilities may seem pedantic, they are nevertheless essential components of an effective learning environment.

THE CLASSROOM. The mathematics classroom should be thought of as a learning laboratory in which a host of multi-sensory aids and manipulative materials are used. However, no room, regardless of either the variety or abundance of physical materials, is in itself a laboratory. The laboratory method in action encompasses much more than just the physical dimensions of the mathematics classroom. In fact, the active involvement of students in problem solving situations will at one time or another utilize a great many school and community facilities. The mathematics laboratory, however, does serve as a primary source for materials, ideas, and learning resources and, as such, should provide a climate in which pupils can pursue a wide variety of mathematical topics.

The classroom can be changed substantially by rearranging pupil desks, chairs and tables. Figure 1 suggests several general arrangements, each of which has both advantages and disadvantages. The teacher selects the classroom arrangement according to the planned learning activities. Changes in learning activities should be accompanied by appropriate changes in the floor plan. Moving furniture can be noisy, so instruct pupils in the techniques of moving desks in a quiet manner. For example, two or more pupils carrying each desk to avoid excessive scraping noise can go a long way toward keeping classroom noise down!

For some schools a large student enrollment and subsequent large number of mathematics classrooms prohibit the establishment of laboratory facilities within each classroom. In such cases the development of a communal room (as shown in Figure 2) can be quite effective. However, the model suggested requires that teachers carefully schedule the use of the room as well as the physical materials in advance. The successful operation of a communal laboratory requires the cooperation of both faculty and students.

Instead of the communal laboratory, some schools provide a variety of mathematics materials within each classroom. In such cases the mathematics laboratory may be a readily accessible alcove in a classroom itself. These classroom laboratories are furnished (to the extent that the budget will allow) with those materials and facilities that are most frequently used. Other equipment and materials are then distributed among the classrooms, with each laboratory providing "homes" for selected items. This model provides considerable flexibility, as students have immediate access to commonly used materials. Furthermore, use of the other facilities can usually be easily arranged by careful scheduling among the teachers. As content emphases change, materials can be exchanged or borrowed from other classrooms within the building.

Two models for mathematics laboratories have been established. A communal laboratory and a laboratory within each classroom. In either case, there are common facilities essential for the success of the laboratory approach.

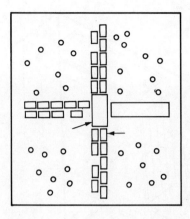

This "cross plan" creates more floor space. The open areas can be used when exploring a variety of different activities.

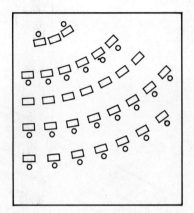

The class directs its attention to one or two people in front of the group. This arrangement can be used to prepare for a lesson or to summarize and evaluate a recently completed activity.

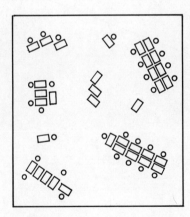

This arrangement allows pupils to work in small group activities and move freely from one work area to another.

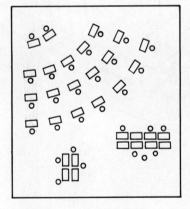

This plan provides for a large group activity while two smaller groups work on projects elsewhere in the room. To minimize distractions, children in the large group face away from the work areas.

Figure 1. Several Classroom Seating Arrangements

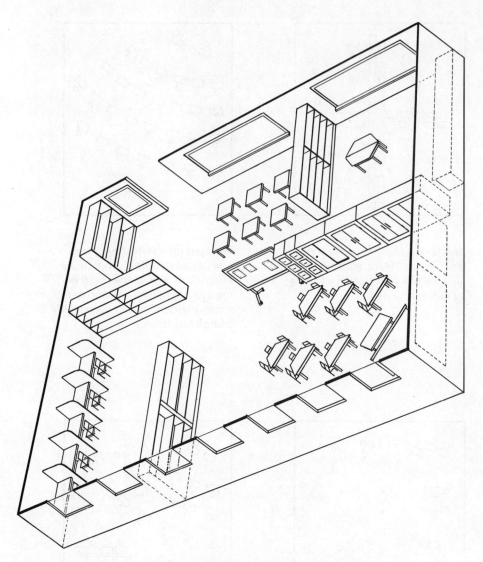

Figure 2. Communal Laboratory

CHALK AND BULLETIN BOARDS. An important ingredient of any mathematics laboratory is an adequate amount of chalk board space. Since pupils will be working on different activities, it is conceivable that a number of groups of pupils will be using the chalkboard simultaneously. As a result, more chalkboard may be required in a mathematics laboratory than in a regular mathematics classroom.

It is suggested that at least two bulletin boards be available. One could be reserved for the teacher to present displays which are an integral part of selected laboratory activities. This bulletin board can be used to present pupils with general problems or puzzles to be solved in the laboratory situation. It can also be used to provide for differentiated assignments of specific problems to certain groups of pupils. A "puzzle of the week" conspicuously placed has, in recent years, served as a motivating device for literally thousands of students.

One bulletin board should be reserved for displaying students' work. Although displays of children's work are often observed in elementary schools, they are found less frequently in the junior high and are rarely found in the senior high classroom. Many secondary teachers feel their pupils have outgrown the need for public display of the results of their efforts. Such a conjecture is generally unfounded. For most people (children and adults alike) there is a great feeling of satisfaction and accomplishment upon seeing one's work displayed. Public display of one's work can be highly motivating and, in fact can often renew one's zeal and enthusiasm for the continued pursuit of learning activities. For specific ideas concerning the effective use of bulletin boards, see *Bulletin Board Displays for Mathematics.*[1]

Graphing daily temperatures

These suggested uses of chalk and bulletin boards are appropriate in either the laboratory or regular classroom setting. The mathematics teacher who is making effective use of the chalkboard and bulletin board is using techniques which are not solely characteristic of a mathematics laboratory.

---

[1] Johnson, D. A., Lund, C. H., and Hamerston, W. D., *Bulletin Board Displays for Mathematics*, Dickenson Publishing Company, Inc. Belmont, California, 1967.

SATELLITE LIBRARY. Most schools have a general library. Whenever possible, provision should be made for satellite libraries to be strategically located throughout the school. If a school has a room designated as the mathematics laboratory, then some library materials might be housed in one section. On the other hand, if each room is considered its own laboratory, then bookcases on coasters or mobile tables could be used to transport reading materials to the individual classrooms. The reading resources should reflect the broad scope of mathematics, and at the same time, provide for wide ranges of reading ability. Toward that end, mathematics readings appropriate for both remedial and enrichment work should be selected. Such material might include textbooks, books about mathematics and mathematicians, special pamphlets and publications, journals and periodicals.*

In most schools the central library should continue to house the majority of mathematics reading materials. The primary objective of the satellite or laboratory library is to make pertinent material in mathematics more readily available to both students and teachers. This also provides a means for displaying new books and current journals. Such reading material is much more likely to be used if it is easily accessible. Students are often "turned on" intellectually by articles, pamphlets or periodicals, their appetites whetted for the learning of more mathematics. Such activity provides a strong rationale for making a satellite library an integral part of a mathematics classroom.

EXHIBIT CASES. Display or exhibit cases provide yet another means of demonstrating the role of mathematics in today's society. A well-lighted exhibit case with adjustable shelves and cork or pegboard backing is quite useful for displaying a great variety of mathematical materials, including mathematical models and other devices prepared by pupils. Projects developed in a mathematics laboratory are often three dimensional. Since it is impossible to exhibit such projects on a bulletin board, a display case serves as an excellent means to protect the project from excessive handling and yet allow it to be easily seen. As with the bulletin board, the opportunity to display classroom projects in this manner serves as strong motivation for many students.

Although display or exhibit cases are best, shelves or long tables may also be used to display materials and projects. Shelves can be adapted to a variety of classroom arrangements and are much less expensive than commercial display cases. Unfortunately, there are disadvantages to using open displays. One limitation of either shelves or tables is that displayed projects are subjected to excessive handling. Furthermore, normal classroom activity frequently prohibits too many tables from being used to display projects for any length of time.

---

* Excellent listings of these materials are found in *Mathematics Library— Elementary and Junior High Schools, High School Mathematics Library* and *A Bibliography of Recreational Mathematics.* Additional information on these National Council of Teachers of Mathematics publications and other references can be found at the end of this Chapter.

STORAGE FACILITIES. A variety of storage space is needed for a mathematics laboratory. Two of the most common problems are lack of adequate storage space and efficient use of space presently available. One basic requirement for selecting any potential storage area is that the stored material must be readily available to both pupils and teachers.

Storage space needs to be provided for equipment, such as projectors, calculators, manipulating and measuring devices, illustrative materials, and construction tools. A teacher's closet, filing cabinet, and desk will provide some storage space. It has been found that storage space for consumable materials, such as lesson or assignment cards, provides the greatest problem in most laboratories. These materials will vary in both size and shape. Furthermore, since the students will be using the materials independently, it is essential that the materials be clearly labeled and directly accessible to students.

Permanent shelves and drawers are ideal for storing laboratory lessons. However, we have found that placing shelving on layers of bricks provides a much more economical and versatile storage area. Placing the bricks between the shelves allows the teacher to adjust the shelf heights to accommodate the kinds of materials being stored. (See Figure 3.) Since this construction is not permanent, it can be altered to serve new needs.

Top view

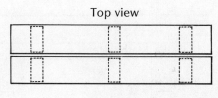

Figure 3A

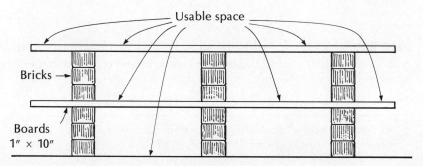

Figure 3

We have found that 1″ × 10″ or 1″ × 12″ shelving works well.* Placing two shelves back to back doubles the amount of storage area, yet consumes only slightly more floor space (Figure 3A). These shelves can also be used to divide the classroom into sections. If groups of students are working on both sides of such a partition this back to back arrangement of the shelves allows pupils to select lessons from either side.

Partitions can easily be made for the shelves by carefully arranging the bricks. Figure 4 shows how the bricks can be moved about on the shelves to divide the usable space into different parts. This is particularly helpful when groups of lessons need to be either clustered together or separated. Lessons can be arranged in a sequence on the shelves which can be labeled. (See Figure 4.) This greatly reduces the chances of students obtaining the wrong lessons.

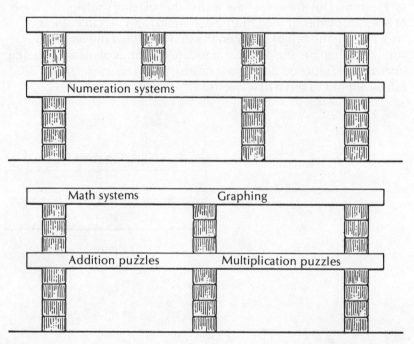

Figure 4.   Usable Space Can be Changed by a Different Placement of the Bricks or Other Supports.

OTHER LABORATORY FURNISHINGS. Standard furniture in a mathematics laboratory should be similar in some respects to that in the general classroom. The ideal mathematics classroom furniture is flat

---

*   As "not so wealthy" graduate students both authors can recall housing personal libraries in this manner during their years in graduate school. If the boards are stained, such bookshelves can, in fact, be made quite attractive.

topped trapezoidal or rectangular tables* and chairs for each pupil. Depending on the size, two to four students could work at each table. These tables provide a large horizontal working surface well suited to the kind of activities students will be involved in. They should provide ample working space for either group or individual work. In addition, such facilities provide great classroom flexibility, as the tables can be arranged to accommodate various situations. For example, several tables may be placed in a cluster to provide for small groups of pupils working together. On the other hand, the tables can be arranged near the walls of the room to provide a large amount of floor space in the center of the room. Such an arrangement is often ideal for classroom demonstration.

The regular teacher furniture in the classroom should include a large desk, chair, filing cabinet, bookcase and storage cabinet. In addition, mathematics teachers need a demonstration table. In fact, such a demonstration table (because of its greater flexibility) is often preferred over the traditional desk by many mathematics teachers.

Other items that need to be provided in the laboratory include: a tape recorder, radio, television set, adding machines, electronic calculators. There should be adequate electrical outlets for these as well as for computer trainer and/or computer terminals.†

The reader has probably observed the similarities of facilities suggested for a laboratory approach and those generally found, although perhaps in varying degrees, in most well equipped mathematics classrooms. We do not recommend that each school set aside a room for a mathematics laboratory, even though many schools have done this successfully. Neither do we recommend that each mathematics classroom house all the facilities mentioned. There are definite advantages and disadvantages in each approach. Perhaps the main advantage of a centralized mathematics laboratory is that all materials are housed in one location and are readily available. On the other hand, one laboratory requires that teachers schedule the room in advance, which frequently prevents teachers from using the room when they want it.

The individual classroom laboratories have the advantage of allowing students and teachers immediate access to the facilities. This usually requires that classrooms share equipment, and sometimes the inavailability of certain facilities prohibits laboratory activities. However, careful lesson planning can usually minimize the time that a class is without a particular piece of equipment. In either case each school faculty must decide on the arrangement which best suits its particular needs.

Happily, this use of mathematics laboratories is not an

---

* We realize full well that relatively few classrooms will be equipped with trapezoidal or even rectangular tables. Regular type student desks can be grouped and rearranged as described above with almost as much ease and flexibility.

† Few schools currently have access to a computer. However, plans for remodeling and building should reflect the future as much as possible.

"either-or" issue. In fact, many schools have successfully combined both of these laboratory concepts. That is, they have both a central mathematics laboratory and individual classroom laboratories with fewer pieces of equipment. Such an arrangement allows for a maximum amount of flexibility and is financially realistic for most schools.

The laboratory concepts should by no means be limited to a classroom or special laboratory room. Indeed, to the laboratory oriented mathematics teacher the world constitutes a laboratory in which mathematics is studied. The classroom represents the laboratory headquarters and is furnished with tools, instruments, references, and materials. It provides both space and facilities for conducting experiments and analyzing data. Such a laboratory makes possible stimulating and worthwhile experiences with mathematics and its many application.

STATION APPROACH TO THE MATHEMATICS LABORATORY

If one is to differentiate instruction within a single classroom, it is important that various activities be organized in such a manner to maximize student learning and, at the same time, minimize class disruption and interference between groups. It is often desirable to physically separate student groups within the classroom. If all pupils assigned to a given group are presented with the same problem situation and are "set apart" from other groups within the classroom, maximum student interaction is promoted and the degree of interference with other groups is kept to a minimum. The "station" is often used to achieve both of these ends. For our purposes, a station can be defined simply as a designated section of the classroom containing all materials (both manipulative and written*) needed for the experiment or task. A possible arrangement is illustrated in Figure 5.

By varying the materials located at various stations throughout the room, and by assigning groups to different locations, it is possible to have a wide variety of mathematical activities in progress simultaneously. Obviously if each station is a duplication of every other one, each group will be involved in basically the same activities. Even under these conditions however, it is unlikely that each group will go about completing the tasks in precisely the same manner. Individual problems are likely to arise within group A that are somewhat unique to that group and need to be handled differently than those of groups B, C, D, etc. This is to be expected, since students, are likely to bring their individual backgrounds to the problem at hand. The station approach lends itself very nicely to such "individualization" within a given problem situation.

Students can move routinely from station to station or groups can be redefined at regular intervals and assigned to stations designed

---

* Hardware and Software are discussed in Chapter 5.

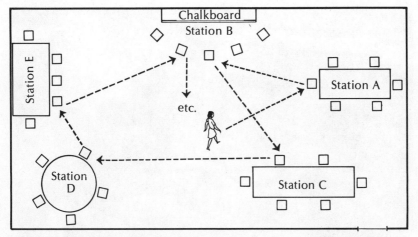

Chalkboard
Station B

Station E

Station A

etc.

Station D

Station C

. . . . *Denotes possible path of Classroom Teacher*

Figure 5

to meet their specific needs. If students are given a choice of selecting their station, the teacher can quickly determine which ones have special appeal. If these more popular stations are examined for their characteristics (i.e. mathematical games, experiments, worksheets, etc.), patterns are likely to emerge. This information is useful both in the designing of future stations and in helping choose commercially available materials for appropriate additions to the laboratory setting.

It is generally desirable to have one extra station available to be used by individuals who complete their assignments early. This extra station need not be concerned with the same theme as the others, and might simply be a game table or individual study center. It is also desirable that the materials and/or mathematical theme(s) of this extra station be changed at regular time intervals.

Some examples of station activities

The classroom teacher utilizing such an approach must assume a role quite different from the traditional "dispenser of information." Under these conditions (especially where stations vary as to mathematical activity) it is not possible to approach the class as a whole. Rather, the teacher becomes a facilitator; one who "roams" from station to station, offering advice, counsel, and encouragement where needed, and generally overseeing classroom activities. When assistance is not needed, it is not given. This frees the teacher to spend his time where it is needed most; that is, with groups requiring special help of some kind.

When first implemented, the station approach is likely to cause some confusion. Don't be dismayed, and stick to your guns. This approach is modifying several time-honored educational doctrines and the student will feel a bit uncomfortable until he begins to understand that he must begin to assume responsibility for his own learning and responsibility for his own actions. With time you will find students responding positively (and in a more organized fashion) to this organizational scheme. The stations approach has been used effectively in elementary and secondary classrooms and we believe it can work effectively in your classroom. Give it a try; you won't be sorry.

SELECTED LEARNING ACTIVITIES

1.   (Group or Class Project) Select and visit a classroom. Take an inventory of the facilities such as moveable furniture, chalkboard and bulletin board space, media equipment and storage area.
   a.   Use this inventory to determine additional needs, if any, for the classroom.
   b.   Sketch a current and then an alternative floor plan for the room. Compare and contrast strengths and weaknesses of these two floor plans.

2.   Design a classroom laboratory capable of accommodating a wide variety of simultaneous activities. (Films, audio tapes, an area within which to conduct experiments, individual reading or research, small group instruction, etc) Identify a partner, exchange ideas and develop another floorplan which contains the best thinking of the two of you.

3.   Identify 4 or 5 "crucial" components of a school communal laboratory. Would your communal laboratory look different if the school for which it were designed was for primary grades only? intermediate grades (4–6) only? junior high (7–9) only? senior high only? HOW?

4.   Group or class project. Select a mathematics topic, such as place value. Describe how a "station approach" might be used to help pupils develop this topic. Be sure to identify activities that could be used at these stations. (Hint: If you want ideas that could be used at different stations, consult *Experiences in Mathematical Ideas.*)

Biggs, Edith E. and James R. MacLean. *Freedom to Learn*. Menlo Park, California: Addison-Wesley, 1969.

Current Publications, Washington D.C.: National Council of Teachers of Mathematics.

Emerging Practices in Mathematics Education, 22nd Yearbook. Washington, D.C.: National Council of Teachers of Mathematics, 1954.

Frame, J. S. "Facilities for Secondary School Mathematics." *The Mathematics Teacher*. 57:(1964): 379–391.

Genkins, Elaine K. "A Case for Flexibility in Classroom Instruction." *The Mathematics Teacher*, 63:(1970): 298–300.

Hardgrove, Clarence Ethel and Miller, Herbert F. *Mathematics Library— Elementary and Junior High School*. Washington, D.C.: National Council of Teachers of Mathematics, 1968.

Hillman, Thomas P. and Barbara Sirois. "In the Name of Geometry." *The Mathematics Teacher*, 61:(1968): 264–265.

Instructional Aids in Mathematics, 34th Yearbook. Emil Berger, Ed. Washington, D.C.: National Council of Teachers of Mathematics, 1973.

Johnson, Donovan A. "A Design for a Modern Mathematics Classroom." *The Bulletin of National Association of Secondary School Principals*, 38:(1954): 151–159.

Johnson, Donovan A. and Gerald R. Rising. *Guidelines for Teaching Mathematics*. Belmont, California. Wadsworth Publishing Co., 1967.

Johnson, D. A.; Lund, C. H.; and Hamerston, W. D. *Bulletin Board Displays for Mathematics*. Belmont, California: Dickenson Publishing Co., 1967.

Kidd, Kenneth P.; Shirley S. Myers; and David M. Cilley. *The Laboratory Approach to Mathematics*. Chicago: Science Research Associates, 1970.

Multi-Sensory Aids in the Teaching of Mathematics, 18th Yearbook. Washington, D.C.: National Council of Teachers of Mathematics, 1945.

National Council of Teachers of Mathematics, *Experiences in Mathematical Ideas*, Volumes I and II, Washington D.C.: National Council of Teachers of Mathematics, 1970.

Raab, Joseph A. *Audiovisual Materials in Mathematics*. Washington, D.C.: National Council of Teachers of Mathematics, 1971.

Schaaf, William L. *Recreational Mathematics*, A Bibliography, Vols. I & II. Washington, D.C.: National Council of Teachers of Mathematics, 1970.

Schaaf, William L. *The High School Mathematics Library* 3rd. ed. Washington, D.C.: National Council of Teachers of Mathematics, 1967.

5 INSTRUCTIONAL MATERIALS WITHIN THE LABORATORY

HARDWARE*

INTRODUCTION.   There are really an unlimited number of materials which could be effectively used in helping students learn mathematics. The quantity of different items needed in a particular mathematics laboratory depends on the objectives of the program as well as the desired learning activities. Obviously, as the variety of learning activities grows, the quantity of supportive materials will also increase. The length of most lists of materials suggested for mathematics laboratories often discourages all but the strongest proponents of this mode of learning. Unless one has access to a large budget (which few classroom teachers do) most lists of materials initially seem to be financially prohibitive. Upon closer examination however it is often found that many of the materials can be secured through students, local appeals, or at a very nominal cost.

Asking students to procure certain materials represents an excellent means of directly involving children in planning and developing instructional activities. For example, if your class is studying liquid measure, an appeal to students to supply containers with an equal or some specific volume will no doubt result in a number of different shaped containers. This could easily lead to discussion of the invariance of volume measure (a quart for example) regardless of container shape. Such ideas are crucial and cannot be properly explored vicariously, but only developed through first hand experience.

The students' motivation will probably increase when using materials they themselves have provided. Furthermore, student involvement in securing materials is likely to make them become more aware of the fact that mathematics permeates all parts of their environment.

A mathematics laboratory includes physical materials which

---

* Hardware is defined here as physical equipment used to supplement the textbook and other printed materials in the mathematics classroom.

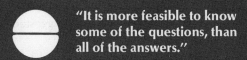

support and promote a wide variety of learning activities. Instructional materials is an amorphous term that encompasses most physical materials found in mathematics laboratories. It would be fruitless to attempt to construct an exhaustive list of either raw materials or tools commonly used in the laboratory setting. Tables 1 and 2 propose materials and tools which have been used successfully and are deemed an important part of the mathematics laboratory facility. Table 1 lists both consumable and nonconsumable materials. The nature of the consumable items makes regular inventory and replacement mandatory. Additional scrutiny of these tables testifies to the general availability of most items. This is very important as most teachers are unfortunately confronted with the task of securing many of their own materials.

Most teachers begin by securing only materials needed for existing laboratory activities. Consequently an increase in the number of activities is usually accompanied by an increase in the quantity and variety of supportive materials. Since materials are not procured until they are needed, this process insures that newly acquired raw materials and tools will be used. The most apparent limitation of this "order as you go" plan is the delay between the time when items are needed and when they are actually obtained. Rarely are teachers— even the most meticulously organized—able to develop long range plans that guarantee the availability of all necessary materials. As a result, some potentially fine learning activities are lost, because students were unable (due to a lack of necessary resources) to "strike while the iron was hot." Few things are more frustrating and disappointing, for both teacher and students than to have an activity aborted for lack of necessary materials.

Tables 1 and 2 identify many necessary materials and tools. These lists should however be viewed only as a starting point in the development of a comprehensive raw material and tool component of the mathematics laboratory. The large number of entries on these lists, combined with the fact that these materials need to be readily available to both students and teachers implies a need for proper storage facilities as well as handling and dispensing procedures.

TABLE 1. Suggested Raw Materials for Laboratory

Adhesives
Adhesive tape
Adhesive wax
Cellophane tape—(colored
    and clear)
Colored plastic tape
Glue
Masking tape
Plaster of paris
Rubber cement
Spray adhesive

Coloring Materials
China Marking pencils
Colored chalk
Colored ink
Crayons
Nylon point pens
Paint
Paint brushes
Spray paint

Fasteners or Binders
Brads
Bolts and wing nuts
Colored rubber bands
Cord
Elastic thread
Hinges
Leather strips
Map tacks or pins
Nails
Pins
Rubber bands
Screws
Staples
Steel strapping
String
Thread
Thumbtacks
Venetian blind cord
Wire
Yarn

Hard and Soft Wood
Balsa wood
Blocks
Colored sticks
Cork panels
Cork tiles
Fiber board
Golf tees
Particle board
Peg board
Pegs
Plywood
Pressed wood
Rulers
Spools
Tack board
Tag board
Tile board
Tongue depressors
Tooth picks
Wooden dowels
Wooden venetian blind slats

Plastics and Paper
Aluminum foil
Blank Cards (different sizes)
Cardboard
Colored corrugated cardboard
Catalogs
Construction paper
Drawing paper
Dress pattern
Egg carton
Gummed letters and figures
Graph paper (rectangular, polar
    coordinate, logarithmic)
Letter stencils
Letters for mounting
Manila paper
Plastic foam
Plastic objects (i.e. toy animals, cars, etc.)
Plastic sheets (plain and colored)
Plastic tubes, bar, and rod
Roadmaps
Sandpaper
Soda straws
Wax paper

TABLE 1.   Suggested Raw Materials for Laboratory (cont.)

Miscellaneous
Balloons
Beads
Billiard balls
Brass strips and bars
Chips
Cotton
Felt
Flannel (assorted colors)
Fruit jar covers
Glass
Levers
Marbles
Materials for making slides
Measuring cups and spoons
Modeling clay
Needles
Perpetual calendars
Pill boxes and bottles
Pipe cleaners
Pocket mirrors

Pulleys
Rings
Sheet metal (brass, copper, tin)
Soap
Steel wire circles
Suction cups
Styro-foam balls and blocks
Tin cans
Tuning fork
Wire mesh

TABLE 2.   Suggested Tools for Laboratory

Tools
Can opener
Carpenter's rule
Carpenter's steel square
Centering square
Crosscut saw
Eyelet punch
Glass cutter
Hack saw
Hammer
Hand coping saw
Hand drill
Knife
Metal shears
Paper cutter
Paper punch
Paper stapler
Plane
Pliers
Plumb line
Scissors
Screw driver
Soldering iron
Straightedge
Tape measure
Vise
Wood rasp

Instruments
Barometer
Clocks
Directional compass
Erector set
Hour glass
Map meter
Pedometer
Scales
Steel tape
Speedometer
Stop watch
Sun dial
Thermometer
Weights

SELECTION CONSIDERATIONS. Classroom teachers of mathematics are witnessing an unprecedented period of proliferation in commercially prepared instructional materials. The future will no doubt witness a continued growth in both quantity and range of quality of these materials. Teachers need to know what materials are available, how to select appropriate items, and how to use them effectively. Just keeping up with current materials is a challenging task. Commercial catalogs list a great variety of available materials; professional journals carry many advertisements which claim their products will provide a panacea for learning a certain mathematical topic; and professional meetings are frequently inundated with exhibits displaying new instructional materials. A list of commercially prepared materials and sources for buying them is provided in Appendix D. Such information should prove valuable and helpful to the classroom teacher.

This influx of newly available materials has precipitated many problems. The wide range of quality found among various materials has made the problem of selection both more difficult and more crucial as the market is flooded with products. It has made it impossible to list all available materials and discuss the merit—or lack of merit—of each. It has been previously suggested that instructional materials should only be used when they serve a specific purpose. Furthermore it should make a unique contribution toward the attainment of specific objectives. Johnson and Rising suggest that the following questions be used in evaluating an instructional device:

1. What purpose does the device serve?
2. Can this device be used by the student to discover a mathematical concept?
3. Is this device to be used by the teacher or by students?
4. Does the device present the idea correctly?
5. Does the model permit the idea to be transferred from the concrete to the abstract?
6. Is the model well designed and sturdy enough to be handled by the students?
7. Is the model large enough so that results are clearly visible?
8. Is the idea which the student derives from the model significant?
9. Does the model do something which cannot be done as well or better than something else?[1]

These questions raise issues to which each teacher should be prepared to respond with regard to each instructional device that is used in the classroom.

There are, of course, many criteria to consider in developing

---

[1] Donovon A. Johnson and Gerald R. Rising, "Using Models to learn Mathematics" Instructional Aids in Mathematics, 34th Yearbook. Washington, D.C.: National Council of Teachers of Mathematics, 1973.

and procuring manipulative materials. In order to simplify discussion, two categories will be considered: pedagogical and physical. The proposed criteria are not exhaustive, nor is any hierarchy of importance suggested by the order in which they are discussed. Although some considerations are more significant than others, the relative importance attached to each criteria should be determined by the teacher. Any final evaluation of instructional materials should be determined by the teacher and should weigh strengths and weaknesses against the educational potential of the devices.

Pedagogically there are many criteria to consider in selecting manipulative materials. One of the most important considerations is whether or not the materials serve the purpose for which they were intended. Furthermore, do these materials do something that could not be done as well or better without them? Since "real" mathematics is abstract in nature, do the materials develop the desired mental imagery?

The following criteria* should be included in any list purporting to identify pedagogical considerations in the selection of instructional materials:

1. The materials should provide a true embodiment for the mathematical concepts or ideas being explored. The materials are intended to provide concrete representations of mathematical principles. Therefore, it is important that, above all else, the materials be mathematically appropriate.

2. The materials should clearly represent the mathematical concept. Some materials embed the concept so deeply that few, if any, pupils extract relevant ideas from their experience with the materials. This problem is further compounded by materials that have extraneous distractions, such as bright colors, which actually serve as a hindrance to concept formation. These experiences result in an "I can't see the forest for the trees" complex. This is, of course, not all bad, as it requires pupils to cull extraneous data, yet in many cases such materials serve more as a deterrent to concept formation than as an aid.

3. The materials should be inherently motivating. There are many factors that ultimately contribute to motivation. Attributes such as attractiveness and simplicity will be discussed under physical criteria. Materials with favorable physical characteristics will frequently stimulate the pupil's imagination and interest.

---

* These criteria are also applicable to demonstration equipment. However, in addition to these criteria, care should be taken to insure that the materials are of sufficient size to be seen by all observers. Furthermore, the teacher should be extremely careful to avoid obstructing pupils' view while demonstrating the device. It is not uncommon for a portion of the class to be screened from a demonstration. In such cases, it is the teacher or seating arrangement, rather than the demonstration, that is at fault. The master teacher anticipates these problems prior to the demonstration and makes plans accordingly.

4. The materials should be multi-purpose. That is, such materials should be appropriate for use in several grade levels as well as for different levels of concept formation. Ideally the materials should also be useful in developing more than a single concept. Such wide applicability is frequently achieved by using a portion or subset of the materials. For example, logic or attribute blocks have a large degree of multi-applicability through the careful selection and use of the various pieces.

5. The materials should provide a basis for abstraction. Nothing can be more distracting than pieces of a tangram puzzle that do not fit properly, or a balance beam that doesn't quite balance. This underscores the importance of the materials correctly embodying the concept. In addition, caution should be exercised to insure that the concept being developed is commensurate with the level of abstraction needed to form the mental image. Care must also be taken to insure that the level of abstraction is commensurate with the ability of the student to abstract.

6. The materials should provide for individual involvement. That is, each student should have ample opportunity to actually handle the materials. This may be done individually or within a group as circumstances dictate. Such use of manipulative materials utilizes several senses, including visual, aural, tactual, and kinesthetic. In general, the materials should exploit as many senses as possible. Compliance with this generalization is particularly important when designing activities for younger children.

Physical criteria are important, since much of the information available to teachers (i.e. commercial catalogs and brochures) describe physical features of the materials. A careful scrutiny of these suggested physical criteria would be helpful in initially screening manipulative materials. Among the physical characteristics to consider in selecting manipulative materials are:

1. Durability—The device must be strong enough to withstand normal use and handling by children. If and when maintenance is needed, it would be readily available at a reasonable cost.

2. Attractiveness—The materials should appeal to the child's natural curiosity and his desire for action. Materials in themselves should not divert attention from the central concepts being developed. Nevertheless there are certain qualities such as aesthetically pleasing design, precision of construction and a durable, smooth and perhaps colorful finish which are desirable.

3. Simplicity—The degree of complexity is of course a function of the concept being developed and of the children involved, but generally the materials should be simple to operate and manipulate. Although the materials may lend themselves to a host of complex and challenging ideas (i.e. the attribute or logic blocks) they should be simple to use. In an effort to construct and use simple devices, there is the inherent danger of oversimplifying or misrepresenting the concept. In all cases care must be taken to insure that the device properly embodies the mathematical concept. In addi-

tion, the design of materials should not require time-consuming, mundane chores such as distributing, collecting, and keeping an extensive inventory record of a large number of items.

4. Suitable size—The materials should be designed to accommodate children's physical competencies and thus be easily manipulated. Storage is an important consideration directly related to size. Each device should be capable of being stored in a reasonable amount of space. The size, as well as the design of the materials, must be carefully planned so as not to produce misconceptions or distorted mental images within the child's mind.

5. Cost—The index used to assess the worth of materials must ultimately weigh their usage against cost. In this context, cost is used in a broad sense. Thus cost must include the initial expenditure, maintenance and replacement charges which would be based on the life expectancy of materials under normal classroom use. Cost is also a function of the additional time required by the teacher to learn about the materials and how to use them effectively.

6. The teacher-related cost is an item of utmost importance. It is not uncommon for someone other than a classroom teacher to order mathematics materials and then expect these materials to be effectively used in the classroom. Of course, this expectation is not often justified! Without proper planning, orientation and preparation, it is ludicrous to expect teachers to effectively use new materials with their pupils. Therefore, any purchase of new materials should be accompanied by a planned program designed to familiarize the teachers with these materials. As a result any cost estimate for instructional materials should reflect the teacher education phase as well as the expenditure for materials.

It would be ideal if instructional materials could meet all of the aforementioned criteria. Finding such materials would be tantamount to finding a "fish that runs fast and flies high." Consequently the search continues. These criteria can, however, be used as a basic guide (see instruments on pages 80 and 81) to the efficient selection of manipulative materials and then expanded to accommodate individual and unique situations.

In addition to these criteria, teachers are often confronted with the dilemma of whether to use commercial or home-made instructional materials. Many of these devices are relatively easy to make and can often be produced by the teacher and/or students. There are many important, intangible by-products, such as additional mathematical insight and increased motivation, directly resulting from these classroom "construction" projects. Nevertheless, one should weigh production costs for homemade materials, including labor, material cost, etc., against the cost of similar commercial products. When the time required to make an instructional device is taken into account, the commercially produced material is usually the least expensive. Quality, of course, is another consideration. Frequently there is a marked quality differential between homemade and commercially produced materials. In such cases the latter is often characterized by better design and construction than homemade materials.

Date_____

Evaluator_____

EVALUATION SCALE FOR A TEACHING AID IN MATHEMATICS

Name of Teaching Aid: _____

Name and Address of Manufacturer or Distributor: _____

_____

Estimated Cost: _____

- -

Instructions: A numerical rating of 0–4 will be used to rate the aid.
4—Excellent; 3—Good; 2—O.K.; 1—Poor; 0—Not Useful.

- -

Rating	Criteria	Comments
— 1.	Material provides concrete representation of mathematical principles.	
— 2.	Relevant ideas from experiences are easily identifiable.	
— 3.	Material is appropriate for several grade levels.	
— 4.	Material is durable. (Would lend itself well to being handled by students)	
— 5.	Material is appropriate for developing different concepts.	
— 6.	Material is simple to operate.	
— 7.	Material appeals to a child's natural curiosity. (Color, packaging, etc.)	
— 8.	Material is appropriately sized for children.	
— 9.	Material can be stored easily when not needed.	
—10.	Material quality is high. (Wood, solid unbreakable plastic, etc.)	
—11.	Material maintenance is low. (Can replacements be obtained?)	

- -

Overall Rating
Recommendation for Grade(s) _____

INSTRUMENT FOR EVALUATING MANIPULATIVE
MATERIALS IN MATHEMATICS

Name of Teaching Aid: _____

Name and Address of Manufacturer of Distributor: _____

_____

Estimated Cost: _____

_ _

Instructions: Circle your rating under each of the following
                criteria.

1. Material provides a true embodiment of the mathematical
    concept(s) or ideas being explored.
    High   Above Average   Average   Low   Not Applicable
    Comment:

2. Material clearly represents the mathematical concept(s).
    High   Above Average   Average   Low   Not Applicable
    Comment:

3. Material would be motivating.
    High   Above Average   Average   Low   Not Applicable
    Comment:

4. Material is multipurpose. (Useable for several grade levels
    and several concepts)
    High   Above Average   Average   Low   Not Applicable
    Comment:

5. Material provides a proper basis for abstraction.
    High   Above Average   Average   Low   Not Applicable
    Comment:

6. Material provides for individual manipulation. (Uses many
    senses)
    High   Above Average   Average   Low   Not Applicable
    Comment:

7. Durability (Including readily available maintenance and partial
    replacement at reasonable cost.)
    High   Above Average   Average   Low   Not Applicable
    Comment:

8. Attractiveness (Pleasing design; precision; finish)
    High   Above Average   Average   Low   Not Applicable
    Comment:

9. Simplicity (In design, use, and distribution within class)
    High   Above Average   Average   Low   Not Applicable
    Comment:

10. Size (Storage; manipulation; prevents misconceptions)
    High   Above Average   Average   Low   Not Applicable
    Comment:

11. Cost (Does it include information and/or inservice training
    to familiarize with materials)
    High   Above Average   Average   Low   Not Applicable
    Comment:

Overall Rating: _____

Recommendation for Grade(s): _____

                            Date _____
                            Evaluator_____

This discussion of commercial versus homemade instruction materials should in no way discourage teachers from constructing devices and/or encouraging their students to do likewise. It is, however, intended to alert teachers to some of the advantages, as well as limitations, of both means of securing materials. Despite the shortcomings of homemade materials, the intangible by-products are many. The authors feel that the active involvement of students in constructing a variety of models or materials is not only highly recommended, but often imperative for the success of the laboratory lesson.

A case in point is construction of polyhedra. Soda straws, pipe cleaners, wood, construction paper or cardboard might be used in constructing different geometric solids. The student will no doubt formulate intuitive notions of faces, edges, vertices, dihedral angles, etc., while constructing these models. This kind of experience cannot be duplicated on other than a first-hand basis. In this situation, learning will be maximized when students have participated in the actual construction of these models. Many excellent books are available that provide ideas and describe ways to construct instructional materials and models in mathematics. Some of these references are listed at the end of this chapter and others may be found in Appendix A.

Children constructing polyhedra

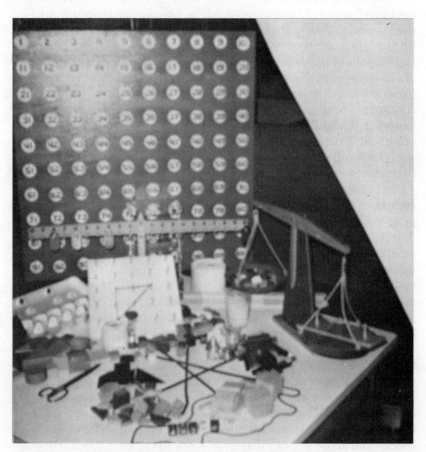

How much of this hardware do you recognize?

SOFTWARE*

INTRODUCTION. It was mentioned earlier that instructional materials encompass a broad domain of items. Thus far the discussion has been limited to hardware, i.e. physical materials that support the learning activity. This hardware does not, in itself, generate learning, but rather provides a basis for concept formation. Learning from these materials requires active involvement by the student. This involvement may take many forms. Perhaps the student reacts to oral and written questions. It may be that the student formulates conjectures and speculations based on his own experience with the materials. Regardless of the nature of student involvement, factors other than hardware must be involved if effective learning is to result. These other factors will ultimately determine the effectiveness or ineffectiveness of the "hard" physical materials.

Other than the teacher, soft instructional materials probably constitute the single most important factor influencing the learning process, since it is from software that learning activities usually emanate.† Textbooks generally dominate the "soft" instructional materials used in most classrooms. However, the challenge to individualize instruction has resulted in the production and use of much additional software in many instructional settings. Consequently in addition to textbooks, many additional books, pamphlets, bulletins, journals, and special publications focusing on selected mathematics topics are commonplace in many schools.

Of all the various types of software developed for mathematics, one of the most valuable outcomes has been the development of task or assignment cards and laboratory lessons. These items represent an important component of "soft" instructional materials and are essential in the implementation of a laboratory approach in mathematics. Subsequent chapters present exemplary laboratory lessons and activities for the elementary and secondary mathematics programs, and are indicative of the important role which software can assume in the development of an activity oriented mathematics program.

TASK, JOB, WORK OR ASSIGNMENT CARDS. For many years the primary source of the learning material in most mathematics classrooms has been the textbook. Provisions for individual differences have been generally limited to differentiated assignments within the regular mathematics text. Since the achievement (as well as ability) range of pupils in a class usually spans several years of growth, it is impossible to adequately provide for individual differences by using

---

* Software is defined here as printed material which generally defines the parameters of the mathematics program.

† It should be noted that hardware and software are equally crucial to the success of the laboratory based program. The software defines activities which are ultimately carried out using available hardware.

a single textbook. Recent years have witnessed a tremendous growth in the variety of mathematics resource materials used by classroom teachers. An array of commercial texts and workbooks compose most of these materials. There has been, however, a noticeable trend toward more individual assignments, independent of regular mathematics books.

Thus teachers are making more curriculum decisions instead of meticulously following a textbook. Individual assignments may be used to provide for skill development, remedial work, enrichment, and motivation, and in this way, supplement as well as complement the current mathematics curriculum. They may also be used to "break the everyday routine" or provide a "change of pace." Special assignments allow teachers to consider student ability, achievement, personality, etc. and the subject matter—topic, level, complexity, etc.—and then select or develop appropriate learning activities. Additional flexibility is provided as the assignments may be made on either an individual or group basis.

Assignments, such as "How many meters wide is our room?" "What is the diameter of this ball?" "Who has the greatest head to waist ratio?" were originally provided orally by teachers. Structuring individual problems and then orally repeating them many times is not only difficult and inefficient, but also very time consuming. Furthermore, such a procedure makes it extremely difficult to provide continuity in assignments. In an effort to develop a sequence of problem solving activities and better provide for pupil differences, teachers began to construct "task" or "assignment" cards. These cards, which are preferable to paper because of their durability, present problems or raise questions at various levels of sophistication. (Figure 1) Some cards, particularly for primary grades and poor readers, portray the problem pictorially or with very few words. Copies of several commercially distributed cards are found on pages 87–89. An annotated list of different task or assignment cards is presented in Appendix C. This listing also provides suggested grade levels, prices, and procurement information.

Cards may be made available to students in several ways. If 4" × 6" or 5" × 7" assignment cards are used, an index box may be effectively used to store and distribute the tasks. On the other hand, some teachers prefer to have the cards displayed on a bulletin board or a pocket chart. These cards might also be displayed on a table that is easily accessible to students. Regardless of the method of distribution, most teachers prefer to store these cards in an index box.

Assignment cards may be individually oriented or addressed to a group. In either case, pursuit of the problems or questions frequently requires the use of selected physical materials. Although the complexity of the assignments vary, the format of the task cards is relatively stable. Generally three basic items are either stated or implied:

1. Statement of problem or posing of question.
2. Identification of any necessary materials.
3. Special instructions, tables or graphs related to the task.

Teacher selecting an assignment card

Assignment cards in a pocket chart

MEASURING WITH STRING

Materials:

1. Piece of string 4–6 inches long
2. Ball of string and scissors

Objectives:

1. Develop the arbitrariness of a unit of measure
2. Provide experience in estimating and measuring

Front of card

MEASURING WITH STRING

Take this piece of string:

1. How many of these pieces do you think would fit end to end around your waist?_____
2. How many actually do?_____
3. How many of these pieces do you think would fit end to end across your desk?_____
4. How many actually do?_____
5. Is the distance across your desk greater than the distance around your waist?_____

Back of card

Figure 1. Exemplary Task or Assignment Card on Measurement

BEST BOUNCE

You will be given 3 lettered balls. Find out how well each ball bounces by dropping each ball from a height of 5 feet. Measure the height that each ball reaches on its 1st, 2nd, 3rd, 4th and 5th bounce.

Record this information in your notebook. Make a graph to show your findings.

---

* This card is one of the *Open Ended Task Cards*, published by Teacher's Exchange of San Francisco.

Drop On

Try to drop six counters on this card.

Show where the
counters land on
a chart like this.

number of counters on card	number of counters off card
2	4

---

* This card is part of *Developmental Math Cards*, Primary Kit-B, by Bates/
Irwin/Hamilton; reprinted by permission of the publishers: Addison-
Wesley (Canada) Limited. c 1970.

Dimes and Pennies

1. Use dimes and pennies
   to show these numbers..

   36 15 47 28
   50 22 30 13

Rule : Use as many dimes as you can
before you use any pennies.

---

* This card is from *Project Mathematics Activity Kit, K-3,* copyrighted by
Holt, Rinehart and Winston of Canada, Ltd. Distributed in the United
States by Winston Press, Minneapolis, Minnesota.

Assignment cards are ideally suited for the mathematics laboratory. Their ability to provide for individual differences is only limited by the variety of cards available. As cards are acquired, they should initially be classified by mathematical topic. Within each topic, care should be taken to identify and provide for different levels of conceptual maturation. It is also necessary to provide for individual differences within each of the various stages of concept development. Thus, assignment cards should be organized in both a "Parallel" and "Series" arrangement. (Figure 2) In the former, a mathematics topic is explored through many different activities, providing for multiembodiment of the concept at a fixed level of development. The series arrangement requires that a continuous sequence of activities be available to provide for conceptual development at successively more difficult levels of sophistication.

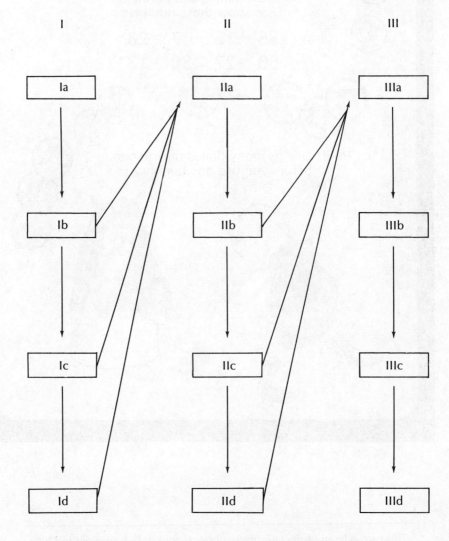

Figure 2. Possible Arrangement of Series and Parallel Activities: A Model for Future Curriculum Development.

Figure 2 illustrates a typical arrangement of these lessons in a "series" (denoted by I, II, and III) each of which contains four "parallel" lessons (denoted by Ia, Ib, Ic, Id, IIa, etc.) This flowchart suggests the sequence that pupils might follow in progressing through the activities. For example, a pupil who (in the judgment of the teacher) has formulated the concept from Ia and Ib will be directed to IIa.* Pupils who have yet to develop the concept after completing Ib, would proceed to Ic. If the concept is grasped after completing Ic, pupils could move directly to IIa, otherwise they would begin Id. This type of arrangement gives each child several "different" opportunities to develop the concept. It also provides for individual differences as pupils are directed to the next activity in the "series" once the concept has been formulated. There is, of course, no absolute number of "parallel" activities required. However, three or four different embodiments of a given concept are usually considered adequate.

It should be clear that time is an important consideration in the series/parallel format. The time required to complete a set of lessons as illustrated in Figure 2 may range from several days to several years. Parallel lessons to develop a particular concept may be completed in one meeting or spread over several days and series lessons may directly follow these parallel lessons. On the other hand there may be other topics or ideas that need to be developed before the next series lesson. In fact a considerable amount of time (weeks, months or even years) may pass before subsequent series lessons are appropriate. The Numeration Systems lessons presented in Chapter 7 are a case in point. The study of the ideas in this series of lessons would normally be spread over several years in the elementary school.

The model proposed in Figure 2 suggests that series and parallel activities are mutually exclusive. Although there are activities that fit this model well, (see Chapter 6 and 7) there are also many activities which require a series and parallel movement simultaneously. For instance, a different parallel lesson could require the pupil to operate at a higher cognitive stage and therefore increase the series level. A case in point is presented in Figure 3, which shows how the model illustrated in Figure 2 might be applied to multiplication.

---

* Some teachers may feel that the concept has been adequately formulated after completing 1a. Therefore, they would recommend the child to move directly to IIa. Although this decision is left to the teacher, the authors feel that abstraction and generalization are often difficult and even foolish when based only on a single experience. Consequently, we have selected a model that assures pupil experience with the concept in at least two different embodiments.

Three different interpretations of multiplication are proposed. Each of these interpretations is progressive in that they require pupils to operate at a higher cognitive level. They are followed by exemplary activities (in both a series and parallel arrangement) that will contribute toward the development of multiplication. Figure 3 presents only a skeletal outline of the actual activities to be used. While only one example is given for each of the above (Ia, for example), additional activities and experiences would be necessary for each child to develop the concept. These different interpretations of multiplication will normally be developed over a period of several years. Thus within each of the interpretations much time will be spent exploring additional properties, such as commutativity and associativity, as well as developing basic skills. The series/parallel format for developing curricular materials is explored more fully in Chapters 6 and 7.

Different commercial companies have developed task cards focusing on many mathematical topics and units of study. (See Appendix C) Most teachers find that commercial cards provide a useful nucleus about which additional activities can be developed. As with other commercial materials, the laboratory teacher must frequently revise, modify, add and delete activities to make them appropriate for his own situation. It of course takes a long time to develop a wide range of cards for a mathematics topic. It should be recognized that the acquisition and development of a system of task or assignment cards is a process of continuous revision and expansion. Just as a teacher must always be alert for new ideas and ways to improve teaching, so must the laboratory teacher be alert for new ideas and ways to improve this form of instructional materials.

In summary, the task or assignment cards are characterized by short problem solving activities focused on selected mathematical topics. Although the format of the cards varies according to pupils and activities, the primary objective is the active involvement of individual pupils in genuine problem solving situations. The tasks usually include student summarization and communication of his findings. This type of experience is designed to promote student involvement in the analysis and synthesis levels of thought operations which are considered to be more sophisticated than mere knowledge acquisition, and thereby more desirable from a pedagogical point of view.

Figure 3.   Example of Series and Parallel Activities

Mathematical Topic—Multiplication

Progressive Interpretations of Multiplication

I   Union of Equivalent Disjoint Sets

II   Cartesian Product

III   Algorithmic

Series and Parallel arrangements within the above interpretations.

I.   Ia   Join 3 sets of 2 buttons each

Ib   Equal jumps on number line i.e. $\overset{\frown\frown\frown}{0\ \ 2\ \ 4\ \ 6}$

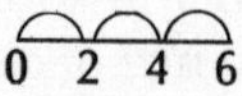

Ic   Rectangular arrays i.e. 3 by 2 array depicted as $\left\{ \begin{matrix} \circ\ \circ \\ \circ\ \circ \\ \circ\ \circ \end{matrix} \right.$

Id   If 3 persons each have 2 shoes, how many shoes in all?

II.   IIa   Problem: I have 3 kinds of ice cream (vanilla, chocolate, strawberry) and 2 kinds of topping (Pistachio and Blueberry). If a sundae is made of 1 kind of ice cream and 1 kind of topping, how many different kinds of sundaes can I make?

IIb   If Nicole has 3 blouses and 2 skirts how many blouse-skirt combinations can she make?

IIc   How many ways can I pair a letter with a number using the following two sets: Set A = (a,b,c); Set B = (1,2).

IId   If Set A has 3 members and Set B has 2 members, how many elements are there in AXB? (AXB is the symbol used to denote the Cartesian Product of Set A and Set B).

III.   IIIa   3·17 interpreted as

$$10 + 7$$
$$\overset{\diagdown 3}{\ }$$
$$\overline{30 + 21 = 51}$$

IIIb*   3·17 interpreted as 3(10 + 7) = 3 · 10 + 3·7 = 30 + 21 = 51

IIIc   3·17 interpreted as

$$\begin{array}{r} 17 \\ \underline{3} \\ 21 \\ \underline{30} \\ 51 \end{array}$$

IIId   3·17 interpreted as

$$\begin{array}{r} 17 \\ \underline{3} \\ 51 \end{array}$$

---

*   Although IIIb is in reality a justification for IIIa (and therefore, using adult logic, should preceed IIIa) children usually experience difficulty with the expanded format of the distributive property (IIIb) if it is considered first. It is, therefore, suggested that the distributive property be considered after the version illustrated in IIIa.

LABORATORY LESSONS.   Development and proper use of assignment cards in both "series" and "parallel" arrangements represents a "bona fide" attempt to enhance individual concept formation. Some activities do not lend themselves as well to assignment cards as they do to other formats. For example, the activity may be too long for effective display on an assignment card. In order to be most meaningful, some activities require pupils to record results or write comments directly on the lesson. If so, a consumable lesson is preferred to a set of cards, which are generally considered reuseable. In such cases as this, laboratory lessons* may be preferred to assignment cards.

Laboratory lessons are also characterized by a carefully developed sequence of questions. Since the lessons may vary in length and complexity, they are (or can be made) appropriate for mathematics students at almost any level.

Laboratory lessons have basically the same characteristics as assignment cards. That is, the lesson is generally problem oriented, and structured so that active participation by the learner is not only encouraged, but required. The primary distinction between assignment cards and laboratory lessons is in the average length of time usually required for the activity. The assignment cards generally require less time for completion than the laboratory lesson. Although it is conceivable that a laboratory lesson may be completed in a matter of a few minutes, most lessons require more time. This time differential has been primarily responsible for assignment cards being identified with the elementary level and laboratory lessons associated with the secondary. In the "ideal" mathematics laboratory, the number of laboratory lessons is very large. In typical laboratories, the variety of available lessons is usually less than·desirable. Regardless of the number of laboratory lessons available, it is the classroom teacher who must decide which lessons are most appropriate for his pupils at any particular time.

---

*   It should be noted that neither of the terms (assignment cards or laboratory lessons) is well defined among mathematics educators. Some say that all laboratory lessons are assignment cards or just a variation of assignment cards. The authors see assignment cards as being short term activities whereas laboratory lessons develop topics which require more time to complete. Any precise delineation between these two terms would be only an academic exercise and would not contribute substantially to this discussion. Although the authors have chosen to discuss these two terms individually, they do not generate mutually exclusive groups. In fact, Figures 4–8 illustrate assignment cards that when taken together form what we would consider to be a laboratory lesson.

Figure 4. 3-Dimensional Problem Solving (3DPS-I)

Count the cubes which have been glued together in each of the
following stacks and record your result. You cannot see all of the
cubes. Assume no cubes are missing in the places you cannot see.

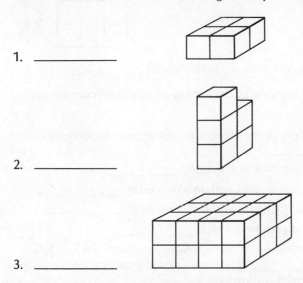

1. _____

2. _____

3. _____

Figure 5. 3-Dimensional Problem Solving (3DPS-II)

Small cubes have been stacked and glued together to form this larger
cube.*

1. How many cubes are in this stack? _____

If the large cube is dropped into a bucket of paint and completely
submerged

2. How many of the smaller cubes are painted on four
   sides? _____

3. How many on three sides? _____

4. How many on two and only two sides? _____

5. How many on one and only one side? _____

6. How many on zero sides? _____

7. What is the sum of your answers to questions 2, 3, 4, 5,
   and 6? _____ How does this compare with your
   answer to question 1? _____

---

* Have cubes available so that pupils may actually construct the larger
   cube, if they wish.

Figure 6.   3-Dimensional Problem Solving (3DPS-III)

Cubes have been stacked and glued together to form this larger cube.

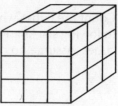

1.   How many cubes are in this stack? _____

If the large cube is dropped into a bucket of paint and completely submerged

2.   How many of the smaller cubes are painted on four sides? _____

3.   How many on three sides? _____

4.   How many on two and only two sides? _____

5.   How many on one and only one side? _____

6.   How many on zero sides? _____

7.   What is the sum of your answers to questions 2, 3, 4, 5, and 6? _____ How does this compare with your answer to question 1? _____

Figure 7.   3-Dimensional Problem Solving (3DSP-IV)

Cubes have been stacked and glued together to form this larger cube.

1.   How many cubes are in this stack? _____

If the large cube is dropped into a bucket of paint and completely submerged

2.   How many of the smaller cubes are painted on four sides? _____

3.   How many on three sides? _____

4.   How many on two and only two sides? _____

5.   How many on one and only one side? _____

6.   How many on zero sides? _____

7.   What is the sum of your answers to questions 2, 3, 4, 5, and 6? _____ How does this compare with your answer to question 1? _____

Figure 8. 3-Dimensional Problem Solving (3DPS-VI)

1. You have solved several problems using the cubes. Using this
information, complete the following Table.*

Length of side of cube	Number of Painted Sides					Total Number of Cubes
	4	3	2	1	0	
2						
3						
4						
5						
6						

2. Suppose your cube were 5 × 5 × 5. Continue the pattern and
record the number of faces with zero, one, two, three and four
sides painted.

3. Do you observe any patterns?_____ If so, complete the
table for a 6 × 6 × 6 cube. If not, sketch or make a cube and
record your responses.

4. Have you really got the idea? If you think so, try to complete the
Table for a 10 × 10 × 10 cube.

CLOSING COMMENTS

In closing this discussion of instructional materials, it should be
clearly understood that the authors are *not suggesting that the entire
mathematics curriculum from K – 12 be laboratory oriented.* We do
strongly feel that the mathematical development of most primary
school children would be greatly enhanced by complete abandon-
ment of traditional textbooks in favor of a STRUCTURED SEQUENCE
(i.e. Series/Parallel lessons) of laboratory oriented mathematical
activities. We also recognize there is *currently an insufficient number
of properly sequenced laboratory oriented activities to justify an
exclusive K – 12 laboratory approach.*

This is not to say that such activities do not exist. In fact, there
are many excellent activities that focus on selected mathematical

---

*   Few students will be able to complete the table for a 10 × 10 × 10 cube
unless some patterns have been identified. Encouraging pupils to keep
track of the factors used in the table aids pattern recognition. For
example, 0, 6, 24, 54 and 96 are the first five correct values for the "1"
column. A pattern is more discernible when these values are written as
0, 6 × 1, 6 × 4, 6 × 9 and 6 × 16.

topics. Unfortunately the number of fully developed and sequenced laboratory activities is quite small when one considers the great breadth and depth of the field of mathematics and the theoretically unlimited number of potential laboratory oriented lessons. Although such activities for *all* mathematical "concepts" generally explored in the schools have yet to be produced, each year witnesses noticeable increases in the quantity and quality of new activities.

Another problem, perhaps more formidable than the development of such activities for all mathematical concepts is that of providing activities which provide for differentiated levels of development. That is, *several different activities each focusing on the same basic mathematical idea should be readily available to all pupils*. These activities should develop the mathematical concept within different physical situations, thereby providing for mathematical and preceptual variability, as discussed in Chapter 3. Such Series and Parallel activities are necessary if one is to provide adequately for individual differences.

These two curricular problems must be overcome if the instructional software is to meet laboratory needs. However, these problems should not be used as an excuse for retention of the status quo. In fact, they should serve as a catalyst, thereby getting more classroom teachers involved in this kind of curriculum development, and subsequent implementation within the classroom. Thus classroom teachers in the future could assume an important role in ultimately determining the quality of instructional materials used in a laboratory approach to the learning of mathematics.

Do not forget, it should be the teacher who either buys instructional materials or recommends their purchase. Prudent control of this buying power could have tremendous impact on the hardware and software produced for mathematics laboratories. Let us hope this "power" is properly used!

SELECTED LEARNING ACTIVITIES

1. What does the phrase "fish that runs fast and flies high" (page 79) suggest about instructional materials?

2. Do you think classroom teachers are more concerned with "hardware" or "software" in teaching mathematics? Defend your position.

3. Visit either an elementary or secondary classroom where mathematics is regularly taught. Take an inventory and then list raw materials and other hardware that you find in this room.

4. Read several task or problem cards. Review cards published by companies such as Science Research Associates and John Wiley or projects such as COLAMBA and Madison Project. Compare the format of the various cards and identify mathematical ideas that could be developed using these cards.

5. Select a mathematical concept, such as fractions. Examine a scope and sequence chart to see how this concept is developed. Then

sketch a series/parallel model that might be followed in presenting the concept.

6.   Small group or class project. Select a mathematical concept. Prepare several task cards related to this topic. Exchange cards with others in your group. Follow the directions on the card to actually complete the activity or perform the task. Jot down questions and/or comments that you have related to the cards and return them to the person who originally prepared the cards. Use this feedback to revise the cards. Once this preliminary revision is completed, use these cards with children to determine if they are effective or if additional revision is needed.

7.   Small group or class project. Suppose a school has $100 to be spent on assignment, task or work cards. You are appointed to a committee that is charged with the responsibility of recommending materials that are to be purchased. Use Appendix C to prepare a list of materials and an accompanying statement briefly explaining why these particular materials are needed.

BIBLIOGRAPHY—CHAPTER 5

Abbott, Janet S. *Learn to Fold and Fold to Learn*. Pasadena, California: Franklin Publishing Company, 1968.

Biggs, Edith E. and James R. Mac Lean. *Freedom to Learn*. Menlo Park, California: Addison-Wesley, 1969.

Chandler, Arnold M., Editor. *Experiences in Mathematical Ideas*, vols. 1 & 2. Washington, D.C.: National Council of Teachers of Mathematics, 1970.

Cundy, H. M., and A. P. Rollett. *Mathematical Models*. New York: Oxford University Press, 1961.

Davidson, Patricia. "An Annotated Bibliography of Suggested Manipulative Devices." *The Arithmetic Teacher*, 15:(1968):509–524.

Denholm, Richard A. *Making and Using Graphs and Nomographs*. Pasadena, California: Franklin Publications, 1968.

Emerging Practices in Mathematics Education, 22nd Yearbook. Washington, D.C.: National Council of Teachers of Mathematics, 1954.

Fitzgerald, William; David Bellamy; Paul Boonstra; John Jones; and William Oosse. *Laboratory Manual for Elementary Mathematics*. Boston: Prindle, Weber and Schmidt, 1973.

Grater, Michael. *Make It With Paper*, London: Miles & Boon, 1961.

Greenes, Carole E.; Robert E. Willcutt; and Mark A. Spikell. *Problem Solving in the Mathematics Laboratory: How To Do It*. Boston: Prindle, Weber and Schmidt, 1972.

Johnson, Donavan A. *Paper Folding for Mathematics Class*. Washington, D.C.: National Council of Teachers of Mathematics, 1957.

Johnson, Donavan A. and Gerald R. Rising. *Guidelines for Teaching Mathematics*. Belmont, California: Wadsworth Publishing Co., 1967.

Kadesch, Robert R. *Math Managerie*, New York: Harper & Row, 1970.

Kennedy, Leonard M. *Models for Mathematics in the Elementary School*. Belmont, California.

Kidd, Kenneth P.; Shirley S. Myers; and David M. Cilley. *The Laboratory Approach to Mathematics*. Chicago: Science Research Associates, 1970.

Krulik, Stephen. *A Handbook of Aids for Teaching Junior-Senior High School Mathematics*. Philadelphia: W. B. Saunders Co., 1971.

Krulik, Stephen. *A Mathematics Laboratory Handbook for Secondary Schools*. Philadelphia: W. B. Saunders Co., 1972.

Laycock, Mary and Gene Watson. *The Fabric of Mathematics*. Hayward, California: Activity Resources Co., 1971.

Multi-Sensory Aids in the Teaching of Mathematics, 18th Yearbook. Washington, D.C.: National Council of Teachers of Mathematics, 1945.

Murray, William D., and Francis I. Rigway. *Paper Folding for Beginners*. Dover Publications, 1960.

Pearcy, J. F. F. *Experiments in Mathematics, Stages 1,2,3*. Boston: Houghton-Mifflin Co. 1967.

Raab, Joseph A. *Audiovisual Materials in Mathematics*. Washington, D.C.: National Council of Teachers of Mathematics, 1971.

Ranucci, Ernest R. *Tessellations and Dissections*. Portland, Maine: J. Weston Walsh, 1970.

Reys, Robert E. "Considerations for Teachers Using Manipulative Materials." *The Arithmetic Teacher*, 18:(1971):551–58.

Steinhaus, H. *Mathematical Snapshots*. New York: Oxford University Press, 1969.

Walter, Marian I. *Boxes, Squares and Other Things*. Washington, D.C.: National Council of Teachers of Mathematics, 1970.

Wenninger, Magnus J. *Polyhedron Models*. Washington, D.C.: National Council of Teachers of Mathematics, 1966.

Williams, C. M. *Sources of Teaching Materials*. Columbus, Ohio: Ohio State University, 1971.

EXAMPLES OF SERIES/
PARALLEL LABORATORY ACTIVITIES

INTRODUCTION

The activities in this chapter are divided into two sections, each illustrating the Series/Parallel approach to mathematics curriculum development. The lessons are designed for use with children, although the reader will find it useful to answer the questions posed. The lessons are activity oriented and will require materials which are readily available or can be easily constructed from odds and ends normally found in the school building.

Most of all, to be successful, these activities require active student involvement. It is appropriate to talk about conclusions *only after* the student has participated in the actual experiment. It is not appropriate to discuss these activities in a "What would have happened if I actually did this experiment?" context. As adults with some knowledge of the mathematical concepts under consideration, we are much more able "to separate the wheat from the chaff" by focusing our attention directly on the relevant issues and are more able to disregard the physical context within which mathematical concepts are embodied or developed. Also as adults with some knowledge of the mathematics under consideration, we are more able to envision the conclusions without necessarily becoming actively involved in the experiment each step along the way. Children cannot afford these luxuries precisely because they do not possess the relevant prior knowledge. It is therefore essential that we as educators provide the concrete experience for children, which will ultimately result in their ability to internalize or abstract the ideas at some later time.

This chapter and the next present laboratory lessons in the Series/Parallel format. The reader will note differences in both the mathematical topics considered and in the methods of presentation. The lessons are quite flexible in nature in that the questions posed are loosely structured. Such questions promote large group discussion, and impose few constraints on the general nature of both acceptable student responses and acceptable student readiness

activities. These lessons for the most part require small group activity,
hypothesis formation, data gathering, data analysis, and group decision making. As a result, one can expect to find students moving
about the classroom interacting with one another, generating a good
deal of noise and periodic "organized confusion." All students completing these lessons cannot be expected to have learned the same
things, since various groups will not have been involved in exactly
the same phases of the various activities. Skillful teacher questioning
and thoughtful discussion following the completion of these activities will, however, serve to develop a similar perspective for all class
members involved.

    The activities in Chapter 7 are more structured. Upon suc-
cessful completion of those activities all students can be expected
to have learned basically the same mathematics, since they have all
been involved in precisely the same activities. Although those lessons
are addressed to individuals, they are most effective when used in
small group situations that allow individuals to interact with the materials in question and then discuss findings or responses with other
members of the group.

    The diversity of approach in these two chapters is the result
of many factors, not the least of which is the partitioning of author
writing responsibilities. This diversity highlights the fact that there
is no one "right" way to develop laboratory activities. The method of
presentation should reflect the philosophy and teaching style of the
person involved. Persons preferring a more organized approach to
the teaching of mathematics would probably prefer the approach
used in Chapter 7. Persons preferring a more open and less structured learning situation (and not overly concerned with classroom
decibel levels) would lean toward the types of lessons presented in
this chapter. Happily it is not an "either-or" situation however, and
both types of activities can be appropriate in a given classroom. The
one essential factor inherent in either approach, however, is that the
children be actively involved both mentally and physically with the
task at hand.

SECTION I
ACTIVITIES WITH PLACE VALUE

The major purpose of this book was not to develop extensive laboratory activities, but rather to develop the rationale for this method of instruction, and to provide specific suggestions pertaining to the development of a laboratory approach in the mathematics classroom. A model for future curriculum development in mathematics was proposed in Chapter 5. You may recall that this model is a direct result of the theory of mathematics learning proposed by Zoltan Dienes.

The laboratory activities proposed in Chapters 5, 6 and 7 were developed to exemplify how the above mentioned model might be used to guide future curriculum development. It is hoped that a second purpose might also be obtained. Activities in these chapters have been developed and tested in the classroom with children, although many of them have been used experimentally with pre- and in-service teachers. If you as a classroom teacher find them (in whole or in part) to be appropriate for your classroom, we encourage you to try them. We think that you'll find the effort personally worthwhile, and, unless our experimental classrooms were quite atypical, you'll find your students will thoroughly enjoy their involvement, and in all probability will learn quite a bit of mathematics as well.

The activities in Section I deal with the concept of place value as exhibited in non-decimal (not base 10) numeration systems.* The variables of weight and length are used to provide parallel embodiments of the mathematical concepts. As you progress you will notice that for each activity dealing with weight there is a parallel (structurally identical to, but differing in physical appearance) activity dealing with length. Each lesson is denoted by either a "W" or an "L" followed by the numeral 2, 3, 4 or 5. The numerals represent the number base under consideration while "W" represents weight and "L" stands for length. Thus activity L-3 deals with the variable length in base 3. Figure 1 (page 111 summarizes the activities in this section and places them within the Series/Parallel Model. If after reading Chapters 6 and 7 you have a clearer concept of what is meant by the Series/Parallel approach to mathematics curriculum development one of our major objectives will have been achieved.

---

*   You'll find a more complete coverage of this topic in Chapter 7. The method of presentation here is quite different however and suggested for use in developing fundamental concepts with younger children.

Objectives:

1. To actively involve students in measuring weight and length to the nearest whole unit.

2. To provide students with the opportunity to reorganize patterns, notice similarity between two seemingly different systems, weights and length (which incidentally are isomorphic to each other and to the system of real numbers) and to draw conclusions about the internal structure of these systems.

3. To expose students to non-decimal numeration systems. (Note that exposure as defined herein does not imply an attempt to develop skill in the computational sense.

Activity W-2

Materials:

1. Given the following weights (*one* each) 1 oz., 2 oz., 4 oz., 8 oz., 16 oz. A collection of weights such as this will be referred to as a set.

   NOTE: Ounces here could be replaced by an arbitrary unit of weight i.e., bottle caps, pennies, small cubes, sugar cubes, etc. The only requirement is that each unit (i.e., sugar cube) has the same weight as every other unit. Thus for sugar cubes, 1 cube would replace the 1 oz. weight, 2 cubes glued together would replace the 2 oz. weight, 4 cubes glued together would replace the 4 oz. weight, etc.

2. Balance Beam. Easily constructed with tinker toys, plastic ruler with hole in center, nail, clips and paper cups.

   NOTE: Be sure that apparatus balances *before* weighing objects. Devices constructed as above are amazingly accurate (and cheap). In the event it does not balance, piece(s) of masking tape can be placed on the light end of the ruler.

Balance beam

3. Each group of two or three students will need a set of weights, a balance beam, a number of objects to be weighed, and a table similar to Table 1.

Student Entry Behaviors: Familiarity with concept of balance beam and in particular that if it balances (ruler parallel to table) the objects in each cup must have the same weight. If students have not had experience with balance beam it is appropriate to introduce them to the principles involved before proceeding further.

Opening or Leading Question: * Using only these weights, can we find (to the nearest unit) the weight of *any* object? Provided, of course, its weight does not exceed 31 ounces, (units)? Try some and see.

Note to the teacher: "31" in the above question can be replaced by "63" if we add a 32 ounce piece to our set of weights. What single piece would we have to add to our set of weights if we wanted to weigh everything up to 127 ounces? (If you answered a "64 ounce piece" you are correct and have obviously noticed a pattern which we would want the children to discover.)

---

* Questions to be asked of students will be identified by §.

§Is there any item which cannot be weighed to the nearest whole unit using only the given set of weights? (Yes, but only those whose weight exceeds the sum of all weights in the set).

§Assume we have 31 objects, each with a different weight, each weighing a whole number of ounces. Show on the following table the pieces you would use to weigh each of them. Put a "1" in the box under a particular weight if you need that weight in order to weigh the object. Example: To weigh object "g" (7 units) I would need one-4 unit weight, one-2 unit weight and one-1 unit weight. (Provide only one or two examples for students, have them complete remainder of table.) It is completely filled out below for your information.

What patterns do you notice in Table 1?

Can you tell how to get from one row to the next? Discuss.

TABLE 1

OBJECT	WEIGHT	16 UNIT WEIGHT	8 UNIT WEIGHT	4 UNIT WEIGHT	2 UNIT WEIGHT	1 UNIT WEIGHT
a	1					1
b	2				1	
c	3				1	1
d	4			1		
e	5			1		1
f	6			1	1	
g	7			1	1	1
h	8		1			
i	9		1			1
j	10		1		1	
k	11		1		1	1
l	12		1	1		
m	13		1	1		1
n	14		1	1	1	
o	15		1	1	1	1
p	16	1				
q	17	1				1
r	18	1			1	1
s	19	1			1	1
t	20	1		1		
u	21	1		1		1
v	22	1		1	1	
w	23	1		1	1	1
x	24	1	1			
y	25	1	1			1
z	26	1	1		1	
aa	27	1	1		1	1
bb	28	1	1	1		
cc	29	1	1	1		1
dd	30	1	1	1	1	
ee	31	1	1	1	1	1

Activity L-2

Perform similar activities using length rather than weight as the variable under consideration. If you were to use paper clips as your unit of length* your set would now include the following five paper clip chains.

1 paper clip long   Paper clips clipped end to end form
2 paper clips   convenient length standards.
4 paper clips   For these lessons disregard the
8 paper clips   fact that the "chains" lose in
16 paper clips   length one thickness of wire in
                       each pair of clips used.

*Students constructing paper
clip chains*

* (Lengths used could be inches, feet, pumpkin seeds or any other uniform standard)

Have students complete a table similar to Table 1 (call this Table 1A) Using paper clips "chains" ask students to measure paper strips (or other units of length).

Activity: Ask students to measure the length of 31 objects (each with a different length, each length being "close to" a whole number of units).

TO THE TEACHER:

Table 1A will be identical to Table 1; except that each time the word "weight" appears in the column headings in Table 1 it shall be replaced by the word "length" in the column heading of Table 1A.

An alternative to providing students with all of the objects to measure would be to have student identify objects in the classroom such that at least one object corresponds to each of the 31 lengths depicted in Table 1A. Lengths can be measured to the nearest paper clip. If this approach is used expect a considerable amount of activity (and disruption) as small groups of students search for the object lengths needed to complete the table. This activity should be a class project where each group of students is assigned only 5 or 6 lengths. This will both decrease the amount of time necessary for such measurements (to roughly 20 or 30 minutes) and serve to promote cooperation within groups. When all (or nearly all) of the objects with specified lengths have been identified, the various groups of students might place their entries on the master chart (Table 1A) which has previously been placed on the chalkboard(s) by the classroom teacher. To avoid confusion at a later time group findings should be OK'd by teacher before they are placed on the chalkboard. Discussion should follow. Through questions students should be encouraged to identify apparent patterns in Table 1A.

§What are the differences between Activities W-2 and L-2?

What is the same? How do Table 1 and Table 1A compare? (They will be identical except that Table 1 deals with weights while Table 1A is concerned with lengths)

Why?

NOTE: This is an example of Dienes Perceptual Variability (Multiple Embodiment) Principle (see Chapter 3 for a more detailed discussion of this principle). Notice the identical mathematical structure (Multiple Embodiment) between the two "suits of clothes" in which the main ideas are embodied (weight in Activity W-2 and length in Activity L-2). When the students realize that both activities are in fact identical, they can be said to have abstracted the concept. (In this case the Base 2 Numeration System).

Activity W-3

Similar to Activity W-2 but using a set with two of each weight. This time each successive piece is 3 times the weight of the previous piece, (i.e., 2-1 oz. weights, 2-3 oz. weights, 2-9 oz. weights, 2-27 oz. weights, etc.) This is the Base 3 Numeration System. Can you label the column headings for the appropriate table for this activity?

Activity L-3

Perform activities similar to W-3 using strips (length) as the variable.

Activity W-4

Similar to Activity W-3 but using a set with three of each weight. This time each successive piece is 4 times the weight of the previous piece (i.e., 3-1 oz. weights, 3-4 oz. weights, 3-16 oz. weights, etc.). (This is the Base 4 Numeration System).

Activity L-4

Perform these activities using the length criterion. (Paper clips or strips of paper or cardboard.)

Activity W-5

Looking back over Activities W-3 and W-4 can you tell what the next step will be if we were to continue the pattern? Base 5, Right! What would the set of weights look like here?

    The sets of Activities (W-2, W-3, W-4, W-5) and (L-2, L-3, L-4, L-5) taken individually are examples of Dienes mathematical variability principle. The mathematical concept here in the final analysis is regrouping (place value) according to some specified pre-determined set of values (i.e., in groups of twos, threes, fours, etc.) The concept of regrouping is not dependent upon the particular agreed upon rule, but rather that one combines and trades smaller parts for larger ones whenever possible within the constraints imposed by the rule in question. Thus, in this case, we are manipulating the mathematical variables (bases) which are *not* an actual part of the concept (place value) itself. It is now possible to relate these activities to the "Series/parallel" format referred to earlier.

Figure 1

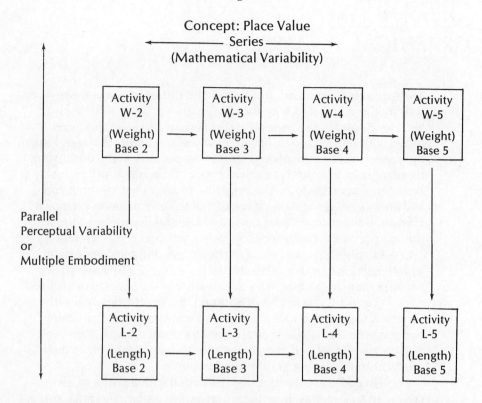

Concept: Place Value
◄———————— Series ————————►
(Mathematical Variability)

Other embodiments of the concept of place value in base two (i.e., place value charts, symbolic algorithms, etc.) would be considered parallel activities and would be correctly placed in the 1st column of Figure 1. If such additional embodiments were then extended to other number bases, they would appear in the appropriate column of the above diagram. That is, the use of place value charts to embody base 5 might be found in Row 3, Column 4.

It will not be necessary for all children to do all of the activities indicated in Figure 1. *You* as a professional teacher must decide which of those activities are essential, and those which are supportive and thereby necessary for only those children who have not acquired the concept satisfactorily from initial experiences. On the other hand, some of these activities might be considered as enrichment and therefore only appropriate for a small portion of the students in your room.

It should be noted, also, that generally the order of activities implied by such a series/parallel matrix is not sacred. Some students might begin with Activity L-2 while others might begin with Activity W-2. It is also conceivable that not all students would start with Activities W-2 or L-2, but instead, might begin with W-3 or L-4. In this case however, it seems reasonable to begin with Activities W-2 or L-2, because they are a bit less cumbersome than other experiences in either the 1st or the 2nd row.

SECTION II
ACTIVITY LESSONS,
GRAPHICAL ANALYSIS

The reader will note both similarities and differences when contrasting these activities with those of Section I. These activities are also developed in the Series/Parallel Format and have also been tested with school-age children. Again we would like to suggest that the reader become involved as much as possible when going through these initially, although realistically your involvement will probably be short of actually doing the activities. In any event we think you will find success in utilizing all or part of these activites in your own classroom situation.* These activities also lend themselves to small group activities. As with Section I, these activities require relatively accessible physical materials of various types, but most importantly, as with Section I (and all laboratory lessons) these activities *require* active student involvement if a full measure of success is to be realized.

Unlike the activities in Section I, these activities deal with a variety of topics from mathematics and science, rather than being concerned with a single topical area. This is not to imply, however, that they are unrelated, for all lend themselves very nicely to mathematical analysis using graphical procedures.

In each case students are presented with a problem, encouraged to generate hypotheses (guesses), gather and analyze data to support or refute their hypotheses, and to draw conclusions. In each case graphical procedures are utilized to aid in the systematic analysis of the problem situation, which is the *major* objective of this series of lessons. This objective is process-oriented, which means, in this case, that factual information gained by students as a result of involvement in these activities is perceived to be less important than the fact that he has rolled up his sleeves, made some guesses, and proceeded to collect relevant information against which to test those guesses. In the process, it is certainly hoped that students will learn something about the irrational number *pi*, or shadows, or motion, or estimation, or plant growth, but keep in mind that such learnings are incidental to the process of carrying out the experiment. It is hard to imagine that students involved in these activities will not learn something about these various topical areas, but precisely what they might learn about shadows, for example, will be very much dependent upon the previous knowledge of small group members, the objects chosen to be measured, the amount of teacher input to group goings-on, and in general, by the nature of the group activities. It is clear that, under these conditions, all children cannot be expected to learn precisely the same material, because each will

---

* Remember that you are the most qualified person to make such a decision as it pertains to your classroom.

*There are
many
embodiments
of coordinate
systems*

have had different experiences relating to that material. This is in contrast to the traditional curriculum where all (or most) students are expected to do precisely the same thing in precisely the same way and at precisely the same time. Incidentally, despite this obsession with conformity, students continue to refuse to learn precisely the same material in precisely the same manner at precisely the same time, which really is not very surprising at all.

The lessons in this section should be fitted to the student, not vice versa. All lessons, therefore, will not be appropriate for all students. Select, modify and/or omit on the basis of what you know about each student and his needs as you perceive them.

It might be useful before proceeding to the activities, to attempt to place them in the Series/Parallel format. (See Figure 2).

Figure 2

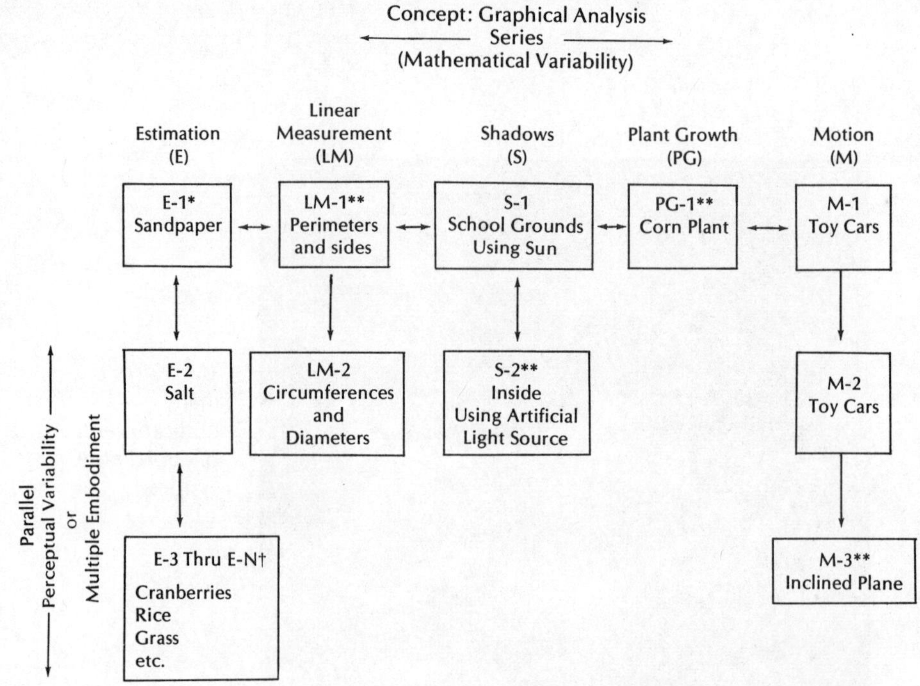

Concept: Graphical Analysis
← ——— Series ——— →
(Mathematical Variability)

Estimation (E)	Linear Measurement (LM)	Shadows (S)	Plant Growth (PG)	Motion (M)
E-1* Sandpaper	LM-1** Perimeters and sides	S-1 School Grounds Using Sun	PG-1** Corn Plant	M-1 Toy Cars
E-2 Salt	LM-2 Circumferences and Diameters	S-2** Inside Using Artificial Light Source		M-2 Toy Cars
E-3 Thru E-N† Cranberries Rice Grass etc.				M-3** Inclined Plane

Parallel — Perceptual Variability — or — Multiple Embodiment

*   For future reference, numbers appearing in activity labels indicate the row number in matrix. Thus E-3 means 3rd activity dealing with estimation procedures, S-2 means 2nd Shadows activity etc.
†   Suggestions for additional parallel activities are presented in this lesson.
    ←——→implies lessons (or set of lessons) interchangeable with respect to order
    ——→implies lessons should be taught in the order indicated by the arrow

Notice a more liberal interpretation of mathematical variability (series) in Figure 2 as compared to that utilized in Figure 1. In Section I, mathematical variability referred to the variation of the number base within which the activities were embedded. In this section, mathematical variability implies actual variation of mathematics and science content areas (estimation, linear measurement, shadows, plant growth, and motion) within the general umbrella of graphical analysis. Parallel activities (perceptual variability) in this section again imply a closely related set of activities where the physical embodiment varies from one activity to the next.

Since each set of parallel activities (columns) deals with topics which are not sequential or logically dependent upon one another, you, as a teacher will have a good deal of flexibility in introducing your students to these activities. The central theme of experimentation/graphical analysis can be found in all activities listed in Figure 2. Two of the columns of activities (estimation. . . . and shadows) can be developed within the context of the station approach as discussed in Chapter 4. Since the individual components of these two sets of parallel activities need not be taught sequentially, the various activities within each parallel unit might constitute the task at a particular station. Taking the estimation unit as an example, one might find four to six stations set up within the classroom, each posing the question "how many," but each asking that question about different sets of physical materials (sand grains, salt granules, cranberries, etc). Individual students would move from station to station until such time that the teacher determines that the particular skill (in this case estimative procedures) has been developed. When this occurs, the student(s) should devote his efforts to other matters. It is likely that some groups will develop such skill before others, thus necessitating the availability of alternate activities.

The station approach was not suggested for the parallel units dealing with linear measurement and motion because, as the arrows in Figure 2 indicate, these activities are sequential in nature and were designed to be taught in a specified order. In the case of the motion unit—M-1, M-2, M-3—relatively large amounts of floor space will be required for the "racetracks." This requirement, coupled with the large amount of student movement, will probably preclude other concurrent activities. The plant growth activities will be spread over approximately a two week period of time (to allow for seed germination) and can easily co-exist with any of the other parallel units.

Students completing the activities indicated by Figure 2 will develop skills in the recording and graphical analysis of experimental data. In addition, they will be continually interacting with their classmates, discussing problem parameters in both small and large group situations. Notice that no grade level designation has been given to any of these activities. We feel that in this case such designation would serve to inhibit their trial in grade levels other than the one designated. If you, the classroom teacher, feel they are appropriate for your students, we hope you will give them a try. We are confident you will find the experience rewarding both for yourself and your students.

ESTIMATION

The lessons in this parallel unit all require the student to answer the question "How many?" In each situation actual counting would be unduly cumbersome. Therefore, the development of a method whereby one can approximate the actual answer using a systematic approach is a necessity. In each case, the student discussion of precisely how to arrive at this "systematic approach" is considered an integral part of the activity. In many instances, students will suggest methods of approach every bit as valid as the ones offered here. When this occurs, by all means, attempt to develop and refine the student suggestions, as such an approach is likely to be more successful than "importing" one of our suggestions. In the event that viable approaches are not forthcoming from your class, you may find our suggestions helpful.

Activity: Estimation #1 (E-1)

OBJECTIVES:

1. To have students utilize graphical procedures to explore and analyze numerical relationships, or patterns emanating from experimentation and measurement.
2. To have students develop systematic procedures for the estimation of seemingly "uncountable" quantities.
3. To introduce prediction by extrapolation and/or interpolation.

MATERIALS:

Very Rough Sandpaper 16 (4 × 4) to 36 (6 × 6) square inches per child. Vary sizes.

Grid—transparent acetate which can be subdivided into smaller square sections. (Overhead transparency blanks cut into six equal areas 4" × 3" work well here.)
Magnifying Glass

Pointer or pencil with sharp point

Using a magnifying glass to count small objects

PROCEDURE:

1. Distribute sheets of sandpaper to each group of 2 to 4 children. Sizes of sandpaper may vary from group to group, but *not* within groups.

2. Ask "How many of largest size grains of sand are there on your sheet of sand paper?" NOTE: It is important that the real answer here is sufficiently large so as to discourage counting. Discuss the reason that counting is not the best approach here (inaccurate, boring, time consuming, etc.). Children should divide into groups of 2 to 4 and attempt to develop a plan for finding the answer to this question.

3. If you determine that various groups have developed reasonable plans encourage them to go ahead and estimate their answers, being ready to explain their procedure to the remainder of the class.

For Emergency Use Only :

1. In the event smaller group discussions are not progressing to your satisfaction, suggest breaking down the large piece of sand paper into smaller areas and counting the number of large grains of sand in the smaller area. You will find the transparent acetate (with perhaps a $\frac{1}{2}$sq. inch drawn on it) and the sharp pencil useful in the counting process.

2. There are at least two ways in which to proceed after finding out how many units of area (say $\frac{1}{2}$ square inch) are contained in the various pieces of sand paper of the various groups. Both are suggested for use here, each being a check on the other.

PROCEDURE 1

Simply multiply: (Number of grains per Unit of Area) × (Number of Units of Area)

PROCEDURE 2

Graph information already possessed, extrapolating (going beyond existing information) to approximate answer. Suggest ordinary $\frac{1}{4}$" graph paper as follows:

In order to generate a larger number of ordered pairs and to insure dispersion of plotted points we suggest that cumulative scores be plotted. Assuming there are 4 students in the group and that each student counted 50 grains of sand per unit, the data table would appear as follows:

Student	Number of Units of Area	No. of Grains of Sand Counted*
John	1	50
John & Mary	1 + 1 = 2	50 + 50 = 100
John & Mary & Mike	1 + 1 + 1 = 3	50 + 50 + 50 = 150
John & Mary & Mike & Mande	1 + 1 + 1 + 1 = 4	50 + 50 + 50 + 50 = 200

The ordered pairs resulting from these data (1, 50), (2, 100) (3, 150) (4, 200) are plotted in Figure (3).

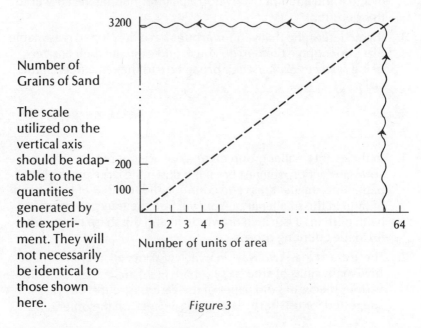

Number of Grains of Sand

The scale utilized on the vertical axis should be adaptable to the quantities generated by the experiment. They will not necessarily be identical to those shown here.

Number of units of area

Figure 3

Questions to pursue with class:

1. Do you notice any pattern? What is it?

2. How could the pattern be extended graphically? (by connecting the points and extending the line segment formed) i.e., see Figure 3 above.

3. Can this graph help you estimate the number of grains of sand there are on your piece of sandpaper? How? (To the teacher: Assuming a 4″ × 4″ piece of sandpaper with $\frac{1}{2}$″ sq. inch units therefore would have 64 units of measure. To find the approximate number of grains of sands here, locate 64 area units on the horizontal axis, proceed vertically to dotted line and then directly left to the vertical axis: the number there represents the number of grains of sand. See the wavy line in Figure 3.)

4. Using the graph can you approximate the number of grains of sand in any piece of your sandpaper if you know its area? (Yes! Procedure similar to that outlined above.)

---

* In reality, the number of grains counted will vary; this will not however affect basic procedure.

5. How does the graphical approximation (Procedure #2) compare with that obtained using Procedure #1? In this case because we assumed a constant student count (i.e. 50 per unit) the results are identical; in reality the results will be close. This is acceptable, however, since we are only charged with the task of estimating, or approximating the true answer.

Activity E-2

OBJECTIVES:

Same as those stated for E-1.

MATERIALS:

1. Box of salt
2. Teaspoons or other *small* container having a relatively constant volume.
3. *Black or other dark colored construction paper. 1 sheet (8½ ×* 11) for each student.
4. Grid.

Grains of salt

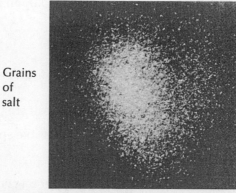

A closer look after dispersion

A still closer look with the aid of a magnifying glass

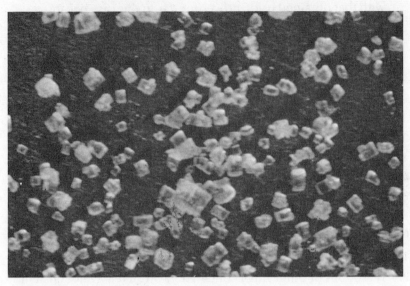

This activity closely parallels E-1. The objectives, procedure and class discussion will be quite similar. In activity E -1 however, grains of sand were conveniently cemented to the sandpaper backing. In this activity, students must develop a procedure for arranging the salt granules so that an agreed upon subset of them may conveniently be counted.

PROCEDURE:

1. Distribute one-eighth to one-fourth teaspoon of salt and construction paper to each student.
2. Ask "How many granules of salt are there in a 'level' teaspoon?"

Discuss with students possible approaches to answering this question.

SHORT DIGRESSION: Once again group discussion is felt to be an integral part of this activity. Answers varying in quality will no doubt be proposed. The large (or small) group exploration of incorrect or unfeasible answers is felt to be every bit as important as the ultimate decision to use the "right" approach. In fact such discussion will often lead to refinement and improvement of the "correct" approach. Keep in mind that it is quite possible that various groups will have suggested different solutions. Such diversity should be encouraged.

This and other lessons in this chapter are examples of process oriented activities.* That is, "how" the student learns

---

* As opposed to "product" or "content oriented" activities. See *The Process of Education* by Jerome Bruner for a more complete discussion of the importance of process in the education of young people.

is every bit as important as "what" he learns. After all who really cares how many grains of salt are actually in a level teaspoon? We certainly don't. The contrived situation however is not trivial. For embedded within it, is one of the most crucial elements of estimation theory. Namely the systematic decomposition of the unmanageable quantity to be estimated into smaller units which can be more easily manipulated. This larger goal is considered worthy of pursuit. The sandpaper, rice, or any other quantity to be estimated are merely vehicles through which to expose the larger idea.

One could apply the same reasoning to any of the activities in Section II of this chapter. The "answers" (product) are not as important as the "method" (process) used to obtain results. Taken as individual entities these activities provide little insofar as important factual knowledge is concerned. Taken individually, they are microcosms of much larger, much more important mathematical ideas. Taken individually none of them are adequately endowed to convey the larger picture to students. Taken as a whole, however, the lessons in Section II of this chapter present a quite different picture. Taken as a whole they provide a relatively wide variety of activities which boil down to a common (and very important) denominator:

1. Exposure to Problem Situation
2. Hypothesis(es) Formation
3. Hypothesis Testing and Evaluation
   (in these cases using graphing procedures)
4. Drawing of Conclusions based on results of above.

You probably recognize this sequence as similar to the "scientific method" spoken of so reverently in science content courses. The fact is that such procedures have played an unbelievably important role in the history of organized knowledge. As such, student exposure to these procedures hardly needs to be justified in terms of its constituting a valid, educational objective. Mathematics and science (and social studies for that matter) provide ideal vehicles through which to involve school age children in such procedures. The laboratory approach to mathematics instruction can be ideally suited for such a purpose.

Taken as a whole, these activities hopefully provide a picture for the student which is more relevant, more complete and more lucid than if these lessons were considered individually or in smaller subsets. The concept that "the whole is greater than the sum of its parts" is a cornerstone of Gestalt psychology, which you may recall from Chapter 3 is the "father" of cognitive psychology, and which you may further recall is the theoretical foundation upon which the laboratory approach as defined in this book is based. It is for the reader to ultimately determine whether or not these activities do, in fact, provide an embodiment of the Gestaltan cornerstone for his particular group of students.

Returning now to our granules of salt:

The method of solution to our problem envisioned here (and by no means the only one) is to:

1. spread the teaspoon of salt evenly on the sheet of black construction paper

2. divide the total area of the construction paper into convenient subsections (say 1 inch squares)

3. determine the approximate number of salt granules per unit of area and

4. extend the result of (3) to the entire area of the construction paper by multiplication and/or graphing procedures similar to that utilized in activity E-1.

If you believe that fractions will present an undue complicating factor for your class, you may want to trim one-half from the construction paper before you begin so that its dimensions will be 8″ × 11″ rather than $8\frac{1}{2}$″ × 11″. It is important that a large sheet of paper be used, however, in order that the salt granules can be sufficiently spread out so as to be conveniently countable.

Questions suggested for activity E-1, with slight modification, are also appropriate here and therefore will not be repeated. In addition, you might want to ask students to approximate the number of granules of salt in a 1 pound box. At this point such a problem becomes almost trivial.

Rather than consume valuable space (and the reader's valuable time) with further detailed descriptions of appropriate further parallel activities E-3, E-4, · · · E-N, let us instead simply indicate that such activities would be similar in mathematical structure but different in physical embodiment or appearance. If one were to manipulate the mathematical variables within the estimation unit, volume or weight might become the variable utilized in the estimation process instead of using area as in E-1 and E-2. As, for example, if one were to estimate the number of cranberries in a 100 pound sack. It would, in fact, be interesting to estimate cranberries having one group use weight, the second group volume and then compare results. The procedures utilized; nature of class or small group discussions, decomposition of whole into manageable subsections,— graphing and extrapolation of collected data, and general nature of conclusions drawn, will be quite similar to those of activities E-1 and E-2.

Other easily obtainable or accessible things that lend themselves to such estimation procedures are:

1. Packages of dried beans such as lentils, pea beans, kidney beans, etc.

2. B-B shot

3. Number of bricks in wall or building

4. Number of "holes" in cyclone fence

5. Number of "loops" in shag rug

6. Number of "holes" in some types of ceiling or floor tile.
7. Number of blades of grass (or weeds) in school lawn. (This involves surveying activities which have additional mathematical payoff)
8. Number of earthworms in designated parcel of land.
9. Number of bacteria in colony.
10. Number of words in book.
11. Number of grains of rice in box, pound jar, etc.*
12. Number of raisins in a loaf of raisin bread by averaging the number of raisins in individual slices (students can literally consume the materials required, when this lesson has been completed).

---

* See *Peas and Particles*, Elementary Science Study, McGraw-Hill Co. for further ideas and sampling procedures.

Estimating number of "holes" in a cyclone fence

Another estimation problem

LINEAR MEASUREMENT

Activity: Linear Measurement #1 (LM-1)

OBJECTIVE:

To have students utilize graphical procedures to explore and analyze numerical relationships or patterns resulting from experimentation and measurement. Specific patterns in this activity and the next are respectively:

1. Relationship between the perimeter of a square and the length of one of its sides. (activity LM-1)
2. Relationship between the circumference of a circle and its diameter. (activity LM-2)

MATERIALS:

Squares of various sizes
Strong tape
Scissors
Large chalkboard of floor grid for each group, approximately 3 or 4 feet square
String or yarn (various colors or thicknesses desirable)

PROCEDURE:

PHASE 1

Divide class into groups of two or three. Using string have students measure the length of the side of a square and its corresponding perimeter. A piece of string should be cut to represent *each* of these measurements. Thus each square will produce two lengths of string, one equal in length to the square's side, the other equal in length to the square's perimeter. Have groups repeat this procedure with 4 or 5 squares. Be sure to have students place a rubber band or tape around each pair of pieces of string to avoid mixing. It is essential that each perimeter be identified with its respective side. Various colored string will simplify the process.

When each group has accumulated 4–6 pairs of pieces of string, have them tape their string directly on a grid similar to Figure 4.

QUESTIONS:

What patterns do you notice? (End points of all perimeter strings will be approximately collinear with point (0, 0) on grid.) i.e., straight edge will pass through (0, 0) and the top point of each perimeter string which has been placed in a vertical position.

If students notice that end points of perimeter strings are not exactly collinear, point out that variations from perfect straight lines are attributable to inaccuracy of measurement.

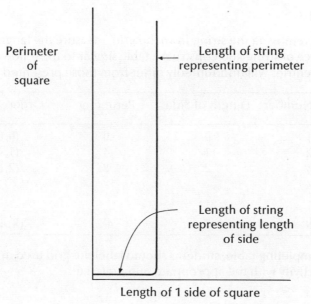

Perimeter of square → Length of string representing perimeter

Length of string representing length of side

Length of 1 side of square

Figure 4

NOTE: Piece of string representing side should coincide with the horizontal axis of the grid. In the above diagram it is drawn above the horizontal or X axis for illustrative purposes only. Do not place numerals on the grid at this point for they (numerals) are not essential to the recognition of the pattern.

When students have placed all (say 5) pairs of lengths of string on the grid it will probably look similar to Figure 5.

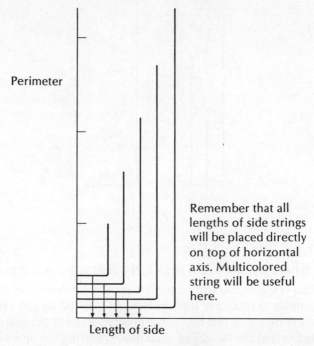

Perimeter

Remember that all lengths of side strings will be placed directly on top of horizontal axis. Multicolored string will be useful here.

Length of side

Figure 5

PHASE II

Without removing the string from the grid, measure the length of each piece of string and construct a table similar to the following. Student entries will undoubtedly differ from those presented here.

Square Number	Length of Side	Perimeter	Ordered Pair
1	0	0	(0,0)
2	1	4	(1,4)
3	2	8	(2,8)
.	.	.	.
.	.	.	.
.	.	.	.
N	X	4X	(X,4X)

After completing table, students should label the grid used in Phase I of this activity with the appropriate numerals, i.e.:

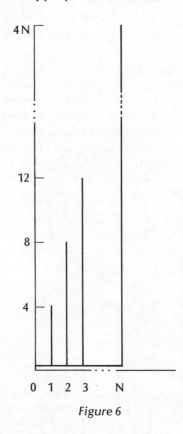

Figure 6

If students have not noticed it before, it is now appropriate to ask:

"What is the numerical relationship between length of side and perimeter of square?" (Perimeter is 4 times the length.) It is easily shown that this relationship is reflected in the graph by noticing that the height of each piece of string is 4 times its distance from the vertical axis. Which simply means that to move

from any point on the line connecting the point (0,0) to the
tips of the vertical pieces of string to any higher point one must
move 4 units upward (Rise) for each unit moved to the right
(Run). For example, to move from A to B,

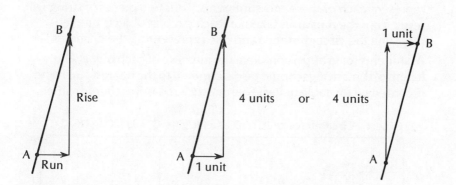

This ratio of Rise/Run is referred to as the slope of the line. (In this
case $\frac{4}{1}$ or $\frac{8}{2}$ or $\frac{2}{1/2}$ or 4). The concept of slope is important mathematically
and can be used to represent a wide variety of physical phenomena.
The reader will note that this concept is utilized repeatedly throughout this series of activities. For example, in the parallel activity to
follow (Linear Measurement #2) the slope of the line generated is an
approximation of the irrational number pi (3.1415 . . .).

Extension of Linear Measurement #1

Perform similar activities with rectangles that are not squares.
Graphing *average* length of sides, i.e.,

$$\frac{L + W}{2} \quad \text{or} \quad \frac{2L + 2W}{4}$$

against its perimeter. Compare results with previous activity
noting similarities and differences.

Activity: Linear Measurement #2 (LM-2)

MATERIALS:

Circular objects of various sizes (disks, plates, balls etc.)
String—various colors if possible
Scissors
Chalkboard grid about 3 feet square

PROCEDURE:

Using string have students measure circumference and diameter of various circular or spherical objects. (For some students it will be an interesting problem to determine a method of finding the diameter of a ball without destroying or bisecting it). String should be cut to represent each of these measurements. Thus two pieces of string will result from the measures taken on each circular object; one representing the circumference and one representing the diameter.

As students (or groups) complete the measure of each object, the length of the 2 strings should be determined to the nearest $\frac{1}{2}$ inch and recorded on the table below. Five or 6 such measures should suffice.

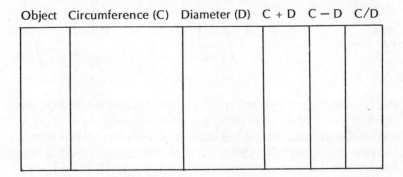

Object	Circumference (C)	Diameter (D)	C + D	C − D	C/D

QUESTIONS:

Do you notice any regularity in the table? (If measures are at all accurate C/D will hover right around 3. Actually the ratio C/D is the mathematical constant π (pi) and has value of 3. 1415 . . . or roughly $3\frac{1}{7}$). Making sure that students keep each diameter string with its corresponding circumference string (various colored string will again be useful) tape the pairs of strings on the board so that the circum-

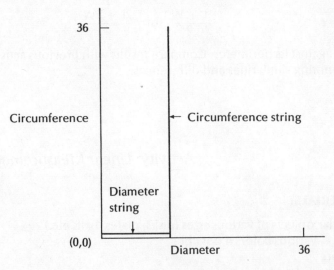

Figure 7

ference strings are vertical and the corresponding diameter string is horizontal, similar to procedure used in LM-1.

With one end at (0, 0), tape a diameter string along the horizontal axis, call the other end point X. Tape its corresponding circumference string in a vertical position, placing one end point at X. See Figure 7.

Repeat this process for several more pairs of string. What do you notice about the intersections on the end points taken as a group? (They fall in a straight line.) Why? Because C/D = constant value, in this case, π (pi), which is approximately $3\frac{1}{7}$. The line through point (0, 0) and these intersections will have a constant slope (C/D) and is therefore equal to π. This can be verified by having students notice that it takes a little more than 3 diameter strings to equal in length the corresponding circumference string.

How could you use the grid on the chalkboard to cut a piece of string so that when it is formed into a circle it will have a diameter of exactly 8 inches? Try it.

Given a string whose length represents the diameter of a sphere or circle, can you, using the graph, determine the corresponding circumference?

After a large group discussion of the chalkboard grid, students might be asked to make several additional measurements, recording their results on their own grids using the numerical (rather than string) format.

SHADOWS:

Activity: Shadows #1 (S-1)

OBJECTIVES:

To provide experience in the recording and analysis of collected data.

To illustrate a functional relationship between two variables. In this case, length of object and its corresponding shadow length.

To provide further opportunity for involvement in the scientific method. That is, identification of relevant contributing factors by carefully controlling various other systems' parameters.

To provide a "real world" example of the concept of ratio and proportion.

MATERIALS:

Various objects with easily discernible lengths. i.e., rulers, sticks, baseball bats, bodies, etc.

Measurement instrument (for length)—ruler, yardstick, tape measure, meter stick.

Light source—Sun.

Playground Area—Recording Sheet, Grid.

PROCEDURE:

On a sunny day take class outside on school grounds. Divide students into groups of two or three. Each group should have some instrument with which to measure length and 4 or 5 objects of different lengths, in addition to a sheet on which to record results such as the following:

Record Sheets		
Date:_____ Time of Day:_____		
Place: *i.e. schoolyard east of building*		
Object	*Length*	*Shadow Length*
Stick		
Yardstick		
Flagpole (see custodian for this measurement)		
Mailbox		
Other		

Have groups of students complete record sheets, the teacher should offer guidance and/or assistance where appropriate. Height of objects are determined by the vertical distance from *ground level* to the highest point.

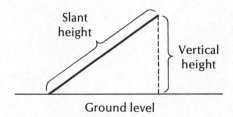

Be sure that objects are held (or implanted in the ground) vertically, as slant height results will differ from vertical height results. Length of shadow is the straightline distance from the base of the object to the farthest tip of the shadow itself.

Return to the classroom and discuss derived measurements. (Remember that all measurement is inexact, so be prepared to discuss variations in students' results.)

For each object there are two measures. It is therefore possible to construct the ordered pair (length of object, length of shadow). Have students plot their ordered pairs on a grid similar to the one below:

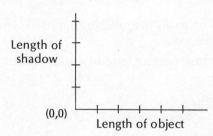

Figure 8

What do you notice?

TO THE TEACHER: Points are collinear with point (0, 0) even though slope or steepness of line is dependent upon time of day measurements are taken. The closer to noon, the smaller the slope (the less steep the line). This happens because shadows are smaller at midday than at any other time. At any given time, however, ratio between length of shadow and length of object is constant, therefore, a straight line will always be generated by plotting derived measurements.

If given the length of an object and the time of day, could you determine the length of its shadow without measuring it directly? (Assuming you already have made a graph of other objects and their shadows at the same time of day.) Yes, by locating the length on the horizontal line (object length) proceeding vertically until the object/shadow line is reached and then reading the shadow length directly on the vertical axis (shadow length). Figure 9 illustrates this approach.

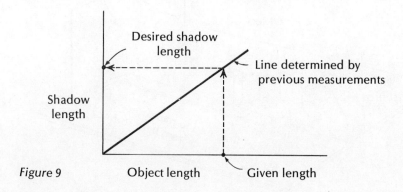

Figure 9

Does one need to know the time of day to answer this question? Yes, if this experiment were previously performed at various times of the day.

If given the length of the shadow of an object and the time of day, is it possible to determine the length of the object? Yes, a procedure similar to that followed above with the order reversed, i.e., see Figure 10.

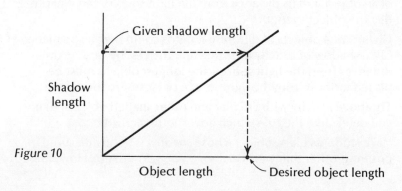

Figure 10

Given a shadow length, is it possible to determine what object made that shadow? No! Many different objects (having the same length) could be responsible for a given shadow. The crucial variable here is object length, not the nature of the object itself.

Activity: Shadows #2 (S-2)

Perform an experiment indoors using overhead or film loop projector, 16 mm or 8 mm projector or slide projector as the light source. Shadows can easily be projected on movie screen, chalkboard, classroom wall, or of course on a floor or table.

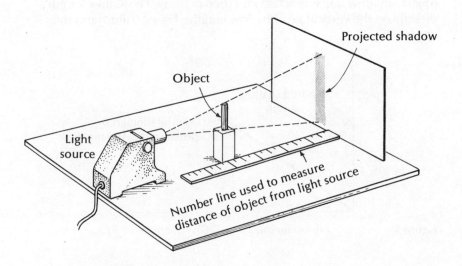

Projected shadow

Object

Light source

Number line used to measure distance of object from light source

Additional Variations on the Same Theme:

a. Vary distance between object and light source. How does this distance affect shadow length?

b. Keeping distance from light source constant and using 3 or 4 objects of different lengths, find their corresponding shadow lengths. Given a 5th object and its length could you predict the length of its shadow? Try it and see. Given the shadow length of an object and its distance from the light source, can you predict the object's length?

c. Given 3 or 4 objects of different lengths could they all be made cast a shadow of exactly the same length? Yes, by varying their distances from the light source, the longest objects must be placed farthest from the light source or closest to the screen.

d. Try above activity with circular and other shaped objects. Can you determine the rule which governs this activity?

e. Have students place object whose shadow is to be measured in positions other than vertical. i.e., ⟋ , ⟍ etc. How does this factor affect length of shadow? This can be done with both the indoor and outdoor phases of the experiment.

f.   Perform outdoor experiment at different times of day, i.e., 10 a.m., 12 noon, and 2 p.m. What do you notice?

g.   Perform outdoor experiments in Fall, Winter and Spring (keeping times of day constant). Careful record keeping over an extended period of time is imperative to the success of this phase of the experiment. If performed conscientiously, however, fruitful discussions of planetary motion and reasons for seasonal change can result. i.e. Keeping time of day and object length constant, why are shadows longest in the winter?

PLANT GROWTH

Activity: Plant Growth #1 (PG-1)

OBJECTIVES:

1.   To provide experience in the recording and analysis of collected data.

2.   To introduce students to interpolation (estimation of the value of an unobserved intermediate value in a known sequence) and extrapolation (estimation of the value of a variable outside of tabulated or observed range).

3.   To illustrate the functional relationship between plant growth and time.

MATERIALS:

Milk Cartons (empty)   1 for each group of 2–3 students.
Soil
Corn Seeds (dried)
Water.
Grid.
Ruler.
Calendar

PROCEDURE: (For Each Group)

Remove one side from milk carton and fill with soil as shown:

Plant dried corn seeds about $\frac{1}{2}$ to 1 inch below the surface on Wednesday or Thursday.

Water until soil is moist, place carton(s) near window, checking daily to see that soil remains moist. Be careful *not* to saturate soil as this will retard germination rate of corn seeds.

Each group of 2 to 3 students should have their own "corn patch."

In 4 or 5 days you will begin to notice sprouts breaking through the soil. Probably Monday or Tuesday of the week following planting of seeds. When this occurs have children measure height of corn plant each day and record results on a grid similar to the one below. In the case of younger children it might be more appropriate to have them cut small strips of paper equal in length to the height of the corn plant each time a measurement is taken. These strips of paper can then be pasted directly on the grid vertically over the appropriate day.

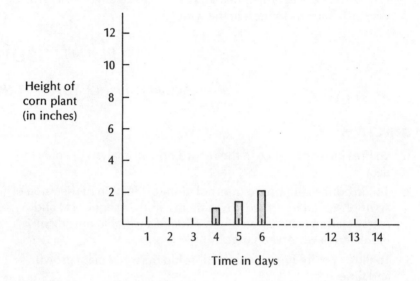

A daily check on the plant's progress

Remember this lesson should begin on a Wednesday or Thursday. As corn seeds grow fast one could expect "results" by the beginning of the following week.

Have each group keep a record of their own plants. These can be compared from time to time (i.e., Who has the tallest plant on day 7, day 10, etc.)

Discuss emerging pattern on grid with class (i.e., plant getting taller with each passing day. Determine how tall (on the average) plants grow each day).

On the Monday following the week when the sprouts began to break the soil surface you will notice that the plant has grown a greater amount than usual. Discuss reasons for this with students (i.e., 3 days have elapsed since last measurement on Friday).

ASK: How tall do you think the plant was on Saturday? On Sunday? (This is interpolation).

ASK: How tall do you think your plant will be in 4 more days? (This is extrapolation).

Students should be encouraged to consult their charts before answering these questions. Discuss.

Extension: Additional Parallel Activities

There are generally three variables which affect seed germination: soil nutrients, sunlight and water.

Experiments similar to that described above can easily be developed to determine the effect each of these have upon the growth of the corn seeds. The basic procedure is to keep two of these variables constant while altering the third i.e.,

1. Same amount of H_2O and sunlight having sand in one carton and top soil in the other.

2. Same soil, same sunlight, one plant is given H_2O the other is not and

3. Same soil, same amount of water, one carton is kept in sunlight the other kept in a dark place.

Variations of this are also possible: i.e., one carton completely saturated with water, one carton soil kept damp, one carton no water. The procedure of attempting to keep the variables constant in order determine the effect of the third upon the system is similar to the basic procedures used in scientific research and therefore is an understanding worthy of pursual.

You will find that students enjoy this experiment.

Question for the teacher: Is this mathematics or science? Or both?

MOTION

Activity: Motion #1 (M-1)

OBJECTIVES:

1. To involve student in active experimentation.
2. To develop skills in data collection, recording and analysis.
3. To encourage hypothesis generation and testing by student.
4. To provide students with an opportunity to quantify aspects of physical reality (in this case a two variable system: time and distance) and to subsequently analyze collected data apart from the physical situation which produced it.
5. To have students determine relative speed of two moving objects by analyzing the graphs representing the motions of those objects.

MATERIALS:

1. Clay 12–16 oz
2. 10–12 sticks (ice cream sticks work fine here) } Each group
3. 50 feet string
4. Stop watch(s). If you are on friendly terms, these can be borrowed from athletic department.
5. 2 self propelled toy cars (*Slow moving* windup cars work very well)
6. Ruler or yardstick
7. Grid

Laying out the track

PROCEDURE:

1. Mark off 20–25 foot runway in 5 foot intervals with a small piece of masking tape.

2. Instruct students as to the use and maintenance of the stop watch —stress careful handling as it is an expensive piece of equipment.

3. To insure that cars will move in a relatively straight line construct two "guide wires" along runway as shown below (8 to 12 inches apart). The string should be about one inch from the floor.

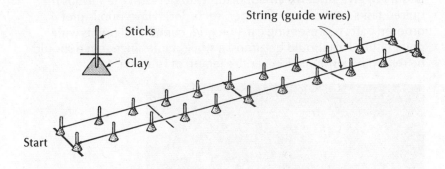

4. Feel free to add as many clay-stick supports as are necessary to stabilize string along runway.

5. Cars should run at different speeds. If both of your cars run alike, place enough clay on top of one of them until there is a noticeable difference in speed.

6. The student(s) with the stop watch should record the time which each car takes to travel (a) 5 feet (b) 10 feet (c) 15 feet and (d) 20 feet. This will require 4 runs for each car unless more than one stop watch is available.

7. Have a student record data collected on chalkboard or record sheet. The data can be conveniently recorded as follows.

Another method of assuring a straight run

Car #1	Time Elapsed (seconds)	Distance	Exemplary Time/Distance Ordered Pairs
These	0	0	(0, 0)
entries are	4	5 feet	(4, 5)
dependent	8	10 feet	(8, 10)
on the speed	12	15 feet	(12, 15)
of the cars used	16	20 feet	(16, 20)

A similar table can be used for car #2.

Be sure to give students an opportunity to perform the various required tasks (starter, time keeper, recorder). Have pupils plot 4 ordered pairs representing car 1 on grid, connect all points with point (0,0). (This should be almost a straight line since speed should be relatively constant along entire length of runway).

The "Big Race"

Sample grid:

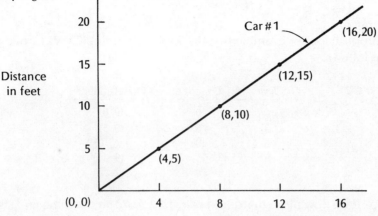

QUESTIONS:

1. Using the graph can you predict how long it would take the car to travel 25 feet? 50 feet? How? Yes to both questions: one merely has to extend the line until it reaches the desired height (distance) then simply determine the related time from the time line (horizontal) below.
   i.e.

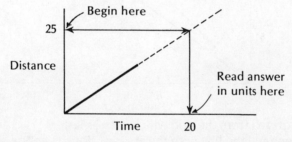

2. Using the grid can you predict the distance traveled in 3 seconds, 15 seconds, etc? Yes. Using a procedure the reverse of the above.
   i.e.

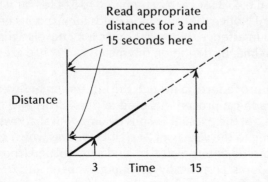

3. Does the graph adequately tell us about what happened with Car #1?

Repeat procedure with Car#2 (In our example Car#2 is faster than Car#1).

Suppose for our example that the Table explaining Car#2 turned out as follows.

Car #2

Time	Distance	Ordered Pair
0	0	(0, 0)
3	5	(3, 5)
6	10	(6,10)
9	15	(9, 15)
12	20	(12, 20)

4. If we now plotted ordered pairs for Car#2 on the same grid, as that used for car#1 we would have

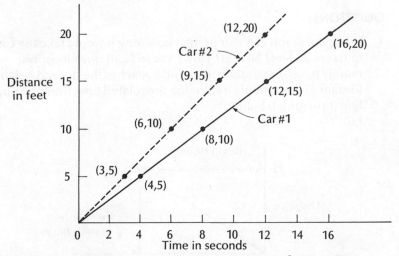

Can we now tell from the grid which car was faster? [Yes car#2 is faster. In general the line with the greatest slope (steepness) represents the faster car.] This is true because ordered pairs (in this case) can be viewed as a ratio between time and distance. The distance/time, (rise/run), ratio for car #2 is consistently greater than the same ratio for car #1. (i.e. 5/3 for car #2 compared to 5/4 for car #1 for the first observation.) Notice the same ratio holds for all points on the respective lines. In essence this means that car #2 traveled the same distances in less time. By definition this means car #2 had greater speed.

Note the similar procedure, (although the interpretations were quite different), of graphical procedures used in this activity and the graphical analysis of the parallel lessons dealing with shadows. There is virtually no limit to the variety of 2 variable systems which can be analyzed, interpreted, interpolated to, and extrapolated from, using graphical procedures. Historically, graphing has been an extremely powerful tool in helping man to quantify and subsequently analyze the world around him. As such it comprises a non-trivial link between mathematics and science and as such can be extremely functional in the laboratory setting.

This activity is identical to M-1 except that cars are now started in different positions and do not necessarily have the same finish point i.e.,

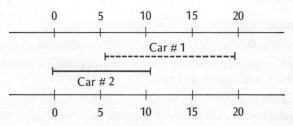

Under these conditions it is not quite so obvious which car has the faster speed. Graphing of ordered pairs generated for each car will again result in two lines (when connected with point 0,0) with different slopes (steepness). Regardless of other factors such as length of time car was in motion or actual distance traveled, the slope of these lines represents the speed of the cars. Using graphical procedures it is now comparatively easy to compare two (or more) moving objects to determine the relative speed by merely comparing the slopes of the lines representing their respective motions or trips.

Considerable experimental work, much of it done by Jean Piaget, indicates that children confuse velocity with speed. Asked to compare the speeds of two toy cars, they say the one reaching the finish line first went faster, regardless of where the two cars were started or how much distance they covered. Until they have had some experience at observing moving objects, they do not consider a car which is overtaking—but far behind—faster than the one being overtaken.

Activity Motion #3 (M-3)

OBJECTIVES:

1. To involve student in active experimentation.
2. To develop skills in data collection, recording and analysis.
3. To encourage hypothesis generation and testing by student. i.e., which ball rolls the fastest?
4. To provide students with an opportunity to quantify some aspect of physical reality (in this case the two variable system: time and distance) and to subsequently analyze collected data irrespective of the physical situation. i.e., by discussing the relative velocity of the various spherical objects represented by the data collected in the experimental situation.

Materials:

1. Inclined Plane—about 20 feet long.*
2. Marble, golf ball, small steel ball, or other spherical objects not more than two inches in diameter.

---

* A very satisfactory inclined plane can be made from cardboard boxes as follows:

1. Cut strips (as long as possible) 6 to 8 inches wide.

2. Carefully fold these strips lengthwise into a "V" shape; this will provide the runway along which the marble, golf ball, and steel ball will travel. Furniture and appliance crates often contain packing with ready made "V" shaped pieces of cardboard.

3. Tape together as many of these sections as are necessary to construct a runway approximately 20 feet long. Tongue depressors or popsicle sticks are helpful in providing stability and support between the various sections. Perhaps you will find longer sticks such as wooden rulers more desirable. These supports should be taped to the bottom sides of the runway so that they will not interfere with the "free roll" of the spherical objects. When the inclined plane is in position, be sure that the cardboard section farthest from the ground is placed on top of the next lower section. This will further reduce interference with the free roll of the marble, golf ball, and steel ball.

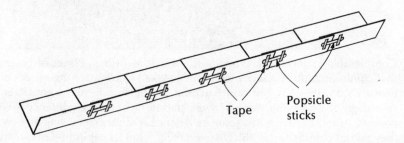

Tape   Popsicle sticks

3. Stop watch(s)—Normally this can be "borrowed" from the athletic department.
4. Chairs, desks, tables, books or other objects to support inclined plane. i.e.,

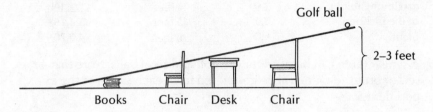

Golf ball

2–3 feet

Books   Chair   Desk   Chair

5. Ruler or Yardstick
6. Grid on which students will record data collected, i,e.,

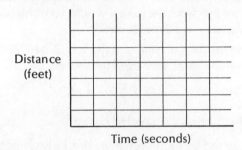

Distance (feet)

Time (seconds)

PROCEDURE:

1. Mark off 5 foot sections on the inclined plane.
2. Elevate one end of the inclined plane so that the highest point is 2 to 3 feet above the floor. Try to avoid as much as possible any "slumping" in the center sections. The various supports should insure a constant "tilt" or slope along the entire length of the runway.
3. Instruct students as to the use and maintenance of the stop watch —Stress careful handling of the watch as it is an expensive piece of equipment.
4. Have students practice rolling the spherical objects down the inclined plane—Note: It should *not* be pushed, if merely placed at the top of the inclined plane and let go, gravity will take it from there.
5. The student with the stop watch should record the time which each of the spherical objects takes to travel, (a) 5 feet, (b) 10 feet, (c) 15 feet and, (d) 20 feet along the inclined plane. This will require 4 runs for each object unless more than one stop watch is available.
6. Have a student record data collected on chalkboard. The data can be conveniently recorded as follows:

OBJECT MARBLE	Time in seconds	Distance	Exemplary Time/Distance Ordered Pairs
These entries will	0	0 feet	(0, 0)
vary with the	1.5	5 feet	(1.5, 5)
maximum height	2.6	10 feet	(2.6, 10)
of the inclined	3.4	15 feet	(3.4,15)
plane.	4.0	20 feet	(4.0,20)

A similar table can be used for the golf and steel balls. Note that 4 sets of ordered pairs will be generated from each phase of the experiment.

1. Be sure to give all students a chance to perform the various tasks, (Roller, Time Keeper, Recorder, etc.)

2. Have pupils plot 4 ordered pairs representing marble on gird. Is there a discernable pattern? Discuss. Using graphed information could we predict how long it would take the ball to roll 25 feet if our runway was 5 feet longer?

3. Do the same for the golf ball and the tennis ball. Plot this information on the same grid used for the marble.

QUESTIONS:

1. Which of the three balls traveled fastest?

2. Would the speed be increased if we used a heavier ball? Try it and see.

3. Was the speed constant from check point to check point? If not, why not?

4. What conclusions can we draw from this experiment?

Extension: Additional Parallel Activities

Replicate the experiment adjusting the inclined plane so that the highest point is 4 or 5 feet from the ground level.

How do the results of the two experiments compare? What conclusions can we now draw as to the relationships between:

a. The nature of the spherical objects rolling down the runway and their respective speeds.

b. The slope of the inclined plane and the speed of the spherical objects.

Why are results here more difficult to predict than those in Activity Motion#1? (Because speed was constant in M#1, in this case acceleration is an additional factor to be considered).

NOTE: This experiment can be replicated using one or more of the sets of commercial racing cars available in most toy stores. The authors have used a variation of "Hot Wheels" and/

or "Johnny Lightening" racing cars with a great deal of success. For those teachers who are not familiar with "Hot Wheels" we suggest that you ask your students. We're sure that they will be able to supply a complete description.

We hope that at this time you are beginning to "get the feel" of a laboratory approach to the learning of mathematics and also an intuitive grasp of the Series/Parallel model for developing laboratory activities. Chapter 7 presents additional activities within the topical area "Numeration Systems".

SUGGESTED REVIEW

The reader will find it useful at this time to review the chapter introduction, the introduction to Sections I and II and the "SHORT DIGRESSION" within the parallel unit dealing with estimation, as these sections in effect present the rationale for laboratory lessons of the type presented in this chapter. Such review will help the reader to place these particular activites within the proper perspective and indicate how they relate to the total laboratory experience.

SELECTED LEARNING EXERCISES

Refer back to the activities in Section I before answering these questions.

1.   Can you imagine how cumbersome such a measuring system would be if it required nine of each weights (or lengths)? It would certainly be more difficult to manipulate. As you know such a system is known as the base 10 system and is the one considered initially (and often exclusively) in the school mathematics program. Do you agree with this practice?

2.   Given the objective: "To have children understand* meaning of place value system," can you develop a rationale for exposing young children initially to the "simpler" systems (i.e., Activities W-2 and L-2) and progressing systematically to the more complex, more cumbersome and inherently more difficult base 10 system?

3.   The previous question implies a different content sequencing in the teaching of place value. This approach would have us consider "simpler" systems with children *prior to* the more detailed consideration of the decimal (base 10) system. Do you find that this suggestion borders on absolute lunacy? Discuss your answer(s) with a friend.

4.   Develop a parallel set of activities dealing with place value in base 10.

Refer back to the activities in Section II before answering these questions.

1.   Name four characteristics which each of the activities in Section II have in common. Discuss these with a friend.

2.   Probability is another topic which lends itself to the development of parallel lessons. In most cases graphs can be utilized to "picture" the results of simple probability experiments. Develop three activities which might constitute such a parallel unit, given the objective "To have students develop an intuitive grasp of the concept 'likelihood of occurrance'".

3.   Develop a card file system whereby series/parallel lessons can be cataloged for future use in your classroom.

4.   Discuss the benefits of such a cataloging system (as developed in 3 above) in a mathematics laboratory (your classroom) utilizing the station approach.

---

*   The behaviorist would have a coronary seeing an objective with such ambiguity as this.

7 MORE SERIES AND PARALLEL ACTIVITY LESSONS

INTRODUCTION

Today there are numerous sources which provide a potpourri of mathematics laboratory activities. These activities may be used for both enrichment and remedial work; they can often be effectively used to complement existing mathematics programs. The "Software" accompanying mathematics laboratory experiments and activities have been embedded in many formats such as books, booklets, pamphlets, workbooks and problem cards. (Some current sources are identified in appendixes A and C.) Since laboratory activities vary considerably in both quality and appropriateness, the classroom teacher must ultimately judge their effectiveness.

A most serious limitation of many of the materials used with teachers (as well as children) is that they generally present "one shot" activities. Little of the software provides a sequence of activities concerned with a major mathematical concept. This section represents an effort to partially fill this void by providing a series of classroom tested* activities for an important mathematical topic. It is hoped that these activities can further serve as a model for additional curriculum development.

These activities provide two different ways of exploring numeration systems through a laboratory approach. Numeration systems were selected because they dominate much of the mathematics that is studied in the elementary and junior high school. This topic also accommodates a broad range of both interest and abilities.

Why should students study different numeration systems? Some proponents of numeration systems argue that students need to study the binary system because of its usage in computers. This is nonsense! Knowledge of base two in and of itself contributes nothing toward the understanding of computers and their operation.

There are other arguments given for studying numeration systems, but few more thought provoking than one cited in a recent

---

* These lessons are for teachers and have been classroom tested with hundreds of preservice and inservice teachers at all grade levels. Although designed for teachers these lessons can, with minor modification, be used with both elementary and secondary students.

survey of intermediate elementary teachers. When asked, "Why do you teach Roman Numerals?" 25 per cent responded "because Roman Numerals are in the basal text!"* This seems hardly a justification for teaching Roman Numerals, or any other numeration system. The fact that numeration systems appear in currently used textbooks may be the only reason why some teachers include this topic in their mathematics programs.

There are of course many different systems of numeration in the mathematics program. Children generally find studying early systems of numeration (Egyptian, Babylonian and Roman) interesting from both a cultural and mathematical point of view. Although the Hindu-Arabic system of numeration with ten as the base occupies most of the curriculum (i.e. from early counting to developing basic skills in fundamental operations of addition, subtraction, multiplication and division) other non-decimal systems are also studied. Presenting other bases through a simple yet meaningful approach has been more difficult than developing early numeration systems. In fact, it is this study of *other bases which have the same properties as our Hindu-Arabic decimal system* that has caused serious problems.

What are the fundamental properties that the Hindu-Arabic decimal system possesses?

1. The system is *additive*.   i.e. $432 = 400 + 30 + 2$

2. The system has *place value*.   i.e. Each digit "3" represents a different value in the numeral 333. The "3" on the left names the number of hundreds; the middle "3" names the number of tens; and the "3" on the right names the number of ones. Thus the position of the digit with respect to the other numerals is important.

3. The system has a symbol for zero.   i.e A symbol "0" which denotes the cardinal number of the empty set.

---

*   R. C. Bradley and N. Wesley Earp, "The Effective Teaching of Roman Numerals in Modern Mathematics Classes, *School Science and Mathematics,* 66:416, May 1966.

Although our Hindu-Arabic decimal system has each of these properties, a study of other number bases reveals that these properties are not unique to the base ten system. The study of these properties in other number bases can lead to greater understanding of the decimal numeration system. For example, "Do patterns result when numbers are multiplied by their respective bases?" Experiences with multiplication in different numeration systems leads to an interesting generalization, of which multiplying by ten in the decimal system is a particular example. We feel the *primary reason for studying non-decimal bases is to develop a greater command and understanding of base ten.* In so doing, the students will practice a variety of skills, although skill development is viewed here as a desirable by-product, rather than the main concern.

Examining different bases that satisfy the properties of additivity, place value, and the existence of a symbol for zero is likely to lead to a better understanding of base ten. Ideas related to topics such as "the role of the base," "place value," "addition and subtraction involving regrouping" and "multiplication involving the base" should also become more clear. The remainder of this chapter presents lessons that exemplify the "series-parallel" model discussed earlier. (See Dienes, Chapters 3 and 5) Two physical embodiments were selected. Nine sequential lessons were subsequently constructed within each of the embodiments. (See Flowchart) The parallel arrangement provides for multiple embodiment or perceptual variability, whereas mathematical variability (different bases) is provided for within each lesson.

Each teacher must decide on the best sequence of the lessons, basing this decision on his own knowledge and experience. This allows for, and in fact requires, addition and/or deletion of specific lessons. It also places the power and responsibility for curricular decisions (i.e. Which of these learning activities are best for each of my students?) where it belongs—in the hands of the classroom teacher.

NUMERATION SYSTEMS VIA MULTIBASE BLOCKS

The following lessons* are designed to provide a foundation for the initial study of numeration systems. These activities only presuppose a knowledge of the fundamental operations in the decimal system. In addition to exploring new systems of numeration, it is anticipated that these lessons will provide a new perspective for understanding base ten. These lessons utilize the Multibase Arithmetic Blocks.†

---

* Remember that *these lessons were written for classroom teachers*—not elementary and/or secondary students. Once you have completed the lessons we think you will want to use many of these activities with your own students. Just be careful about using them in toto!

† Multibase Arithmetic Blocks, including rectangular, triangular, and

FIGURE 1. Flowchart of Series – Parallel Lessons Suggested for Numeration Systems*

SERIES →

PARALLEL →

	BASIC VOCABULARY AND RULES	PLACE VALUE AND ORDER RELATIONSHIPS	EQUIVALENT FORMS IN DIFFERENT BASES	NUMBER AND SUCCESSOR	ADDITION AND SUBTRACTION	MULTIPLICATION AND DIVISION
	NS–1A NS–2A	NS–3A	NS–4A NS–5A	NS–6A	NS–7A	NS–8A NS–9A
	NS–1B NS–2B	NS–3B	NS–4B NS–5B	NS–6B	NS–7B	NS–8B NS–9B

* The cells in this flowchart identify lessons related to the column headings. Each lesson has been coded. Consider the lesson NS–7A. "NS" denotes lessons used to study numeration systems. "A" identifies this lesson as dealing with one perceptual variable, namely the multibase blocks. Thus all NS–A lessons use the multibase blocks to develop numeration systems. The numeral preceeding "A" identifies the ordinal position of the lesson. So NS–7A is the seventh lesson using the multibase blocks to study numeration systems.

The lesson NS–7A provides experiences for developing fundamental addition and subtraction skills in different bases. In this way mathematical variability is provided within each of the perceptual settings. NS–7B is a parallel lesson and reflects the same skills, but is developed within a different physical context. Specific objectives for each of these lessons are also identified.

trapezoidal blocks are available thru Webster/McGraw-Hill 1221 Avenue of the Americas, New York, New York, 10020. The current cost of the revised classroom set (Bases 2, 3, 4 and 5) is $69. The original M.A.B. materials (Bases 3, 4, 5, 6 and 10-rectangular only) are available in classroom size sets from Creative Publications, P.O. 10328, Palo Alto, Calif. 94303. Current cost $16 5.00.

Figure 2. Multibase Strips*

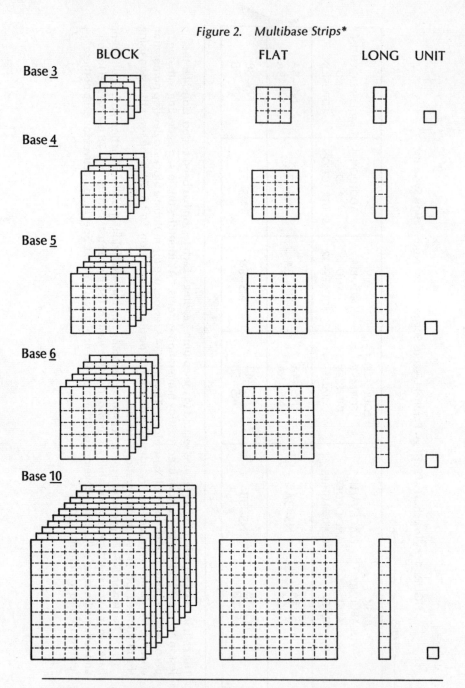

BLOCK FLAT LONG UNIT

Base 3

Base 4

Base 5

Base 6

Base 10

---

* Rectangular strips are illustrated here. Other shapes such as triangles may also be used. For example:

provides a concrete embodiment for the same concept as Base 4 in Figure 2.

The Multibase Arithmetic Blocks (MAB) were developed by Zoltan Dienes to demonstrate the concept of place value. The blocks are generally used with individual or small groups of children who perform tasks identified by assignment cards. This feature permits a good deal of flexibility, and promotes individualized student instruction. Various subsets of the MAB materials deal with number bases other than 10. Each subset consists of units, longs, flats and blocks. The ratio of $\frac{units}{longs}$ and $\frac{longs}{flats}$ and $\frac{flats}{blocks}$ is constant in each subset. This factor permits student study of the concept of place value in several different number bases (Mathematical Variability Principle). The rectangular blocks are illustrated in Figure 3.

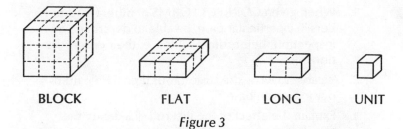

BLOCK   FLAT   LONG   UNIT

Figure 3

If the multibase blocks are not available, similar results can be obtained by using strips of tagboard or construction paper. Use of multibase strips will necessitate some minor revisions in terminology in lessons 1A through 9A. Although any size pieces may be used, we suggest that units be made one half inch square for intermediate children and one inch square for primary children. (The dimensions of the unit determine the dimensions of all other pieces.) The multibase strips are illustrated in Figure 2.

The lessons in this chapter are sequential in nature and should therefore be completed consecutively. The importance of carefully reading the lesson and then following directions cannot be over emphasized. Liberal use of the wood is encouraged but not required. You will find that some lessons move more rapidly than others. Do not worry about progressing too slowly.

Each lesson is preceeded by several specific objectives. These objectives alert you to the nature of the lessons and provide guidelines for minimal levels of competency. You should review these objectives at the end of each lesson and check to see that each of them has been achieved.

OVERVIEW OF LESSON OBJECTIVES

Upon successful completion of the respective lessons you should:

NS-1A

1. Be able to state relationships among UNITS, LONGS, FLATS, BLOCKS, LONG BLOCKS, etc. for a given box.

2. Be able to make fair trades using wood within a given box*.

NS-2A

1. When given a pile of wood, be able to make all necessary trades, write the result or CORRECT FORM, and read it properly.

2. When working in a particular base, determine which symbols could appear in CORRECT FORM.

3. Be able to demonstrate why there is one and only one CORRECT FORM for any number.

NS-3A

1. When given CORRECT FORMS of different numbers in a particular base, be able to determine by inspection their ordinal value, i.e. their order relationship.

2. Be able to write the largest four digit whole number in a given base.

3. Explain the effect of the first roll of a die on the eventual outcome in the Place Value Game.

---

* The term "Box" is frequently used in these lessons. Shoeboxes are ideal for storing the multibase blocks and also separating the different bases. Thus Box 3 refers to a shoebox that contains UNITS, LONGS, FLATS and BLOCKS for base three. Although "box" is used, the term is synonymous

Pieces in Box <u>3</u>

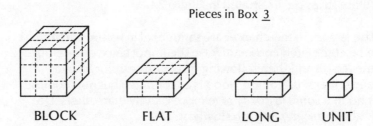

BLOCK   FLAT   LONG   UNIT

with "base" in the context of these lessons. Box <u>4,</u> Box <u>5,</u> and Box <u>6</u> would contain base four, five and six pieces respectively. The actual number of each piece needed in a box is a function of the base. However for these lessons, approximately 20 UNITS are needed together with at least N^2 LONGS, FLATS and BLOCKS, where N is the base. Thus in Box <u>3,</u> 20 UNITS, 9 LONGS, 9 FLATS and 9 BLOCKS provide sufficient material for a group, (4-6), of students. The quantity of material needed also depends on the pupils' age and ability. This quantity of individual pieces generally becomes less essential once fundamental relationships among the pieces are established. The original M.A.B. materials (Bases 3,4,5,6 and 10— rectangular only) are available in classroom size sets from Creative Publications, P.O. 10328, Palo Alto, California 94303. Current cost: $120.00.

NS-4A

1. Be able to determine by inspection the number of UNITS, LONGS, FLATS, BLOCKS, etc. when given the CORRECT FORM.

2. When given a total number of units, be able to determine the CORRECT FORM for a particular base.

3. When given the CORRECT FORM and a particular base, be able to determine the total number of units represented.

NS-5A

1. When given the CORRECT FORM in one base, be able to determine the CORRECT FORM for the same amount of wood (i.e. same total number of units) in another base.

2. When given the CORRECT FORM of two numbers in different bases, be able to determine which has the most units.

NS-6A

1. Be able to write the largest one digit number and its successor.

2. Be able to write the largest two digit number and its successor.

3. Be able to write the largest three digit number and its successor.

4. Be able to write the largest four digit number and its successor.

NS-7A

1. Be able to add and subtract two, three and four digit numbers in particular bases.

2. Be able to determine the base given addition and/or subtraction problems that are solved correctly.

NS-8A

1. Be able to determine the CORRECT FORM of the product in a particular base, given a multiplicand and a one digit multiplier.

2. Be able to determine the CORRECT FORM of a product in a particular base given a multiplicand and the base or power of its base as a multiplier.

NS-9A

1. Be able to solve a division problem by partitioning (separating) a pile of wood into equivalent groups and recording the CORRECT FORM.

2. Be able to determine the CORRECT FORM of a quotient in a particular base, given a dividend and a base or power of the base as a divisor.

NUMERATION SYSTEMS 1A

Upon successful completion of this lesson you should:

1. Be able to state relationships among UNITS, LONGS, FLATS, BLOCKS LONG BLOCKS, etc. for a given box.
2. Be able to make fair trades using wood within a given box.

This is a UNIT

This is a LONG

This is a FLAT

This is a BLOCK

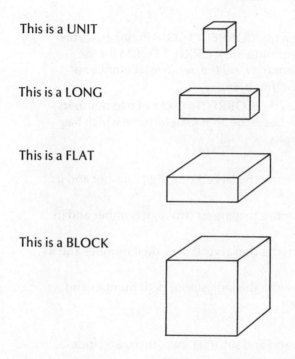

Place value display

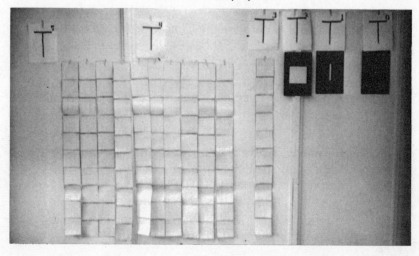

Select Box 3 or another box which you have, and answer all the
questions about that box on pages 157 and 158, then select another
box and start again on page 157. Place your answers in the appropriate cells. Fill in the cells for Box 10 and Box b when you understand how to find the correct responses.

	Box 3	Box __	Box 10	Box b *
1. Construct a LONG out of UNITS. How many UNITS does it take to make a LONG?				
2. Construct a FLAT out of LONGS. How many LONGS does it take to make a FLAT?				
3. Construct a BLOCK out of FLATS. How many FLATS does it take to make a BLOCK?				
4. Construct a LONG out of BLOCKS. We will call this a LONG BLOCK. How many BLOCKS make a LONG BLOCK?				
5. Construct a FLAT out of LONG BLOCKS. We will call this a FLAT BLOCK. How many LONG BLOCKS make a FLAT BLOCK?				

6. What name would you give the wood piece one size larger than
   a FLAT BLOCK? (Take another look at the UNIT-LONG-FLAT-
   BLOCK-LONG BLOCK-FLAT BLOCK sequence.)

   _____ .

7. If you were to construct this wood piece, (see answer to question 6), how
   many FLAT BLOCKS would it take?

Box 3	Box __	Box 10	Box b

8. Do you think this process of building bigger wood pieces would ever
   end? (Assume you have an inexhaustible supply of wood.) _____

* b denotes any arbitrary base.

Suppose we are working with the wood from Box <u>3,</u> then it takes_____
UNITS to make 1 LONG. We can give this process a new name, TRADING.
Using this new name, we would say: We TRADE 3 UNITS for 1 LONG.

Suppose you are in the TRADING business and you have been asked the
following question: (You may use the wood pieces and a trading board to
answer the questions whenever you wish, however be alert for patterns.)

	Box <u>3</u>	Box <u> </u>	Box <u>10</u>	Box <u>b</u>
1. How many UNITS trade for a LONG?				
2. How many UNITS trade for a FLAT?				
3. How many UNITS trade for a BLOCK?				
4. How many UNITS trade for a LONG BLOCK?				
5. How many LONGS trade for a FLAT?				
6. How many LONGS trade for a BLOCK?				
7. How many LONGS trade for a LONG BLOCK?				
8. How many FLATS trade for a BLOCK?				
9. How many FLATS trade for a LONG BLOCK?				
10. How may FLATS trade for a FLAT BLOCK?				

TRADING OR GAME BOARD

	LONG BLOCKS	BLOCKS	FLATS	LONGS	UNITS
Pile 1					
Pile 2					
Pile 3					

*Children
using
a
trading
board*

OPTIONAL LESSON FOR THOSE DESIRING ADDITIONAL PRACTICE IN TRADING

You're in the trading business and it is your job to do as much trading as you possibly can. Fill in this business report.

Today I traded:

Box 3	Box 5	Box b
__UNITS for 1 LONG	__UNITS for 1 LONG	__UNITS for 1 LONG
__LONGS for 1 FLAT	__LONGS for 1 FLAT	__LONGS for 1 FLAT
__FLATS for 1 BLOCK	__FLATS for 1 BLOCK	__FLATS for 1 BLOCK
__BLOCKS for 1 LONG-BLOCK	__BLOCKS for 1 LONG-BLOCK	__BLOCKS for 1 LONG-BLOCK
__LONG-BLOCKS for 1 FLAT-BLOCKS	__LONG-BLOCKS for 1 FLAT-BLOCK	__LONG-BLOCKS for 1 FLAT-BLOCK
__FLAT-BLOCKS for 1 BLOCK-BLOCK	__FLAT-BLOCKS for 1 BLOCK-BLOCK	__FLAT-BLOCKS for 1 BLOCK-BLOCK
__UNITS for 1 FLAT	__UNITS for 1 FLAT	__UNITS for 1 FLAT
__LONGS for 1 BLOCK	__LONGS for 1 BLOCK	__LONGS for 1 BLOCK
__FLATS for 1 LONG-BLOCK	__FLATS for 1 LONG-BLOCK	__FLATS for 1 LONG-BLOCK

NUMERATION SYSTEMS 2A

Upon successful completion of this lesson you should:

1. When given a pile of wood, be able to make all necessary trades, write the result or CORRECT FORM, and read it properly.

2. When working in a particular base, determine which symbols could appear in CORRECT FORM.

3. Be able to demonstrate why there is one and only one CORRECT FORM for any number.

Now take from Box 3 2 BLOCKS, 4 FLATS, 2 LONGS, and 3 UNITS. Write this as follows:

BLOCKS	FLATS	LONGS	UNITS
2	4	2	3

Before Trade

Now trade so your pile looks like this:

BLOCKS	FLATS	LONGS	UNITS
3	1	3	0

After Trade (1)

1. Now count the total number of wood pieces. (i.e. The number of BLOCKS + the number of FLATS + the number of LONGS + the number of UNITS.) Write the total number of pieces here _____ .

Now trade again so your pile looks like this:

LONG BLOCKS	BLOCKS	FLATS	LONGS	UNITS
1	0	2	0	0

After Trade (2)

CORRECT FORM
1 0 2 0 0

2. Again count the total number of wood pieces in this pile. (i.e. The number of LONG BLOCKS + the number of BLOCKS, etc.) Write the total number of pieces here_____ .

3. Are there more or fewer pieces in "after trade (1)" than in "after trade (2)"?_____ .

4. Could you make any more trades to get a fewer number of pieces?_____ .

If all possible trades* have been made we will represent the result in CORRECT FORM. Check the CORRECT FORM in the example, i.e., 10200 This is read "'one'-'zero'-'two'-'zero'-'zero' Box three" or "'one'-'zero'-'two'-'zero'-'zero' Base three." Always read the COR-RECT FORM of a number in this way.

Let's try some more trading:

Take some wood from Box 4 or Box 5. You don't have to use all the pieces in the box. Count the pieces of each size that you took out and write them as before.

BLOCKS	FLATS	LONGS	UNITS

Total number of
pieces Before Trade
_____

Now TRADE—making all possible trades before writing your answer.

LONG BLOCKS	BLOCKS	FLATS	LONGS	UNITS

Total number of pieces
After All Trades_____

1. Fill in the CORRECT FORM._____
2. *Could* a "6" appear in this CORRECT FORM?_____
3. *Could* "8" appear in any CORRECT FORM for Box 4 or Box 5? _____
4. Circle the symbols which <u>could</u> appear in this CORRECT FORM:
   0 1 2 3 4 5 6 7 8 9

---

* Have you noticed that all of our trading has been upward, i.e. trading for fewer pieces. For example in Box 3, we would trade three LONGS for a FLAT rather than for nine UNITS. It will be assumed throughout the first six lessons that *whenever we trade, we always trade for fewer pieces.*

Let's try some more trading.

Take some wood from Box 3. Now count the number of pieces of each size and write the corresponding symbol in the appropriate cell.

LONG BLOCKS	BLOCKS	FLATS	LONGS	UNITS	
					Total number of pieces Before Trade_____

Now TRADE—making all possible trades before writing your answer.

LONG BLOCKS	BLOCKS	FLATS	LONGS	UNITS	
					Total number of pieces After All Trades_____

1. Fill in the CORRECT FORM._____

2. Could a "4" appear in the CORRECT FORM above? Explain._____
   _____ .

3. Circle the symbols that <u>could</u> appear in this CORRECT FORM:
   0   1   2   3   4   5   6   7   8   9.

4. Do you have the same amount of wood (i.e. same total number of units) before and after you traded?_____

5. How does the total number of pieces "before trade" compare with the total number of pieces "after all trades", i.e., are there more or fewer pieces?_____ .

Hopefully you have discovered that "after all trades" never has more pieces than "before trade." In fact "after all trades" has the LEAST number of wood pieces possible. We will call the number represented by the LEAST number of wood pieces the CORRECT FORM. From now on we will only use the CORRECT FORM to symbolically represent the amount of wood in a pile.

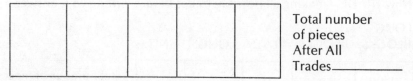

Correct form for these pieces is 1 2 4 3

Now try this problem:

The teacher gave Sandy, David, and Mike each four BLOCKS, eight FLATS, five LONGS, four UNITS. She took all the wood pieces from Box <u>3.</u> The teacher told the children to do all the necessary trading and to write the CORRECT FORM. Here are the answers the children reported:

	LONG BLOCKS	BLOCKS	FLATS	LONGS	UNITS
SANDY'S		5	6	3	1

	LONG BLOCKS	BLOCKS	FLATS	LONGS	UNITS
DAVID'S	1	2	2	0	1

	LONG BLOCKS	BLOCKS	FLATS	LONGS	UNITS
MIKE'S	2	1	1	0	1

	LONG BLOCKS	BLOCKS	FLATS	LONGS	UNITS
Your Answer					

1. Who has the CORRECT FORM? _____
2. Which of the children made an unfair trade? _____
3. Is there another way to write the CORRECT FORM of your answer? _____

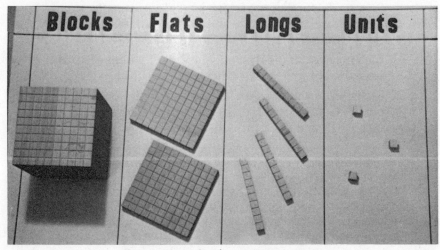

Correct form for these pieces is 1 2 4 3

NUMERATION SYSTEMS 3A

PLACE VALUE GAME

Upon successful completion of this lesson you should:

1. When given CORRECT FORMS of different numbers in a particular base, be able to determine by inspection their ordinal value, i.e. their order relationship.

2. Be able to write the largest four digit whole number in a given base.

3. Explain the effect of the first roll of a die on the eventual outcome in the Place Value Game.

Diagrammatic Pattern of Models for Dice*
Used in the Place Game

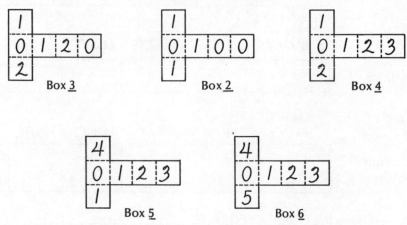

Box <u>3</u>   Box <u>2</u>   Box <u>4</u>

Box <u>5</u>   Box <u>6</u>

Directions:

1. Play this game in small groups (approximately 2–4)

2. Select a box.

3. Leave all the wood of one base in the box.

4. Select a die for that box. (Check the diagrammatic patterns for the appropriate die to use and use only this die.) Be sure to observe and then record the numerals on the die.

5. Each player rolls the die and selects the number of BLOCKS, from the box as is shown on the die.

6. Each player rolls the die a second time to determine the number of FLATS selected. The third throw determines the number of LONGS, and the fourth the number of UNITS.

---

* Wooden or plastic cubes with appropriate numerals can serves as dice. Small cubes of foam rubber or styrofoam will reduce the noise level. Dice may also be made from construction paper using the above patterns. There is nothing sacred about the arrangement of the numerals on the die. However it is essential that the die contain only numerals appropriate for that base.

7. Record the number of BLOCKS, FLATS, LONGS, and UNITS in the appropriate boxes opposite the player's name.

8. For each game record (left to right) under column <u>Summary</u> the number of BLOCKS, FLATS, LONGS, and UNITS.

9. After each person has had four throws, the winner will be the player with the most wood. i.e. If each player's wood was placed on a scale, the pile which weighed the most would win.

10. Place a check (√) to the right of the winner.

Box_____   Numerals on die_____

Player Name	1st throw BLOCKS	2nd throw FLATS	3rd throw LONGS	4th throw UNITS	Summary
					_ _ _ _
					_ _ _ _
					_ _ _ _

Box_____   Numerals on die_____

Player Name	1st throw BLOCKS	2nd throw FLATS	3rd throw LONGS	4th throw UNITS	Summary
					_ _ _ _
					_ _ _ _
					_ _ _ _

Box_____   Numerals on die_____

Player Name	1st throw BLOCKS	2nd throw FLATS	3rd throw LONGS	4th throw UNITS	Summary
					_ _ _ _
					_ _ _ _
					_ _ _ _
					_ _ _ _

Answer the questions below if the summary for a game is:

Name	Summary
Scott	2 3 1 4
Sally	2 3 4 1
Kelly	3 2 4 1
Timmy	3 2 1 4

1. Who has the most wood?_____
2. Who has the least wood?_____
3. Bob says, "You can't determine who has the most wood until you know the box that was used." Martha says you can. Who is right and why?_____
   _____

After you have played this game several times, with different boxes, consider the following questions:

1. What faces on the die would result in the maximum amount of wood for four throws?
   a. In Box 3?   __ __ __ __
   b. In Box 4?   __ __ __ __
   c. In Box 5?   __ __ __ __
   d. In Box 6?   __ __ __ __

2. If the players roll different numbers on the first throw, can you tell which person will win the game_____ Why?

3. If the players roll the same number on the first roll, what can you say about the person who rolls the greatest number on the second roll?

4. Suppose you used Box 4 wood with a conventional die. Would your answer to question 2 be the same? Explain or cite an example.

Suppose the game is changed so that the first throw determines the number of UNITS, the second throw determines the number of LONGS, the third throw determines the number of FLATS and the fourth throw the number of BLOCKS.

5. How does this affect the outcome of the game?

Playing the place value game with Box 3

Comparing towers to determine the winner!

NUMERATION SYSTEMS 4A

Upon successful completion of this lesson you should:

1. Be able to determine by inspection the number of UNITS, LONGS, FLATS, BLOCKS, etc. when given the CORRECT FORM.

2. When given a total number of units, be able to determine the CORRECT FORM for a particular base.

3. When given the CORRECT FORM and a particular base, be able to determine the total number of units represented.

Suppose you have one LONG and two UNITS. We have agreed to record these pieces in a particular manner.

BLOCKS	FLATS	LONGS	UNITS	Summary
		1	2	12

The symbol to represent one LONG and two UNITS would be 12. The value in the Summary column is called the CORRECT FORM of a number.

How many UNITS, LONGS, FLATS, etc. would each of the values in the Summary represent? (Fill in the blanks)

LONG BLOCKS	BLOCKS	FLATS	LONGS	UNITS	Summary
					122
					21
					2
					200
					1001
					102
					10101

If you know the CORRECT FORM can you tell the total number of units represented by 122? What else do you need to know? _____

Determine how many units are represented by each of the following:

	CORRECT FORM After all trades	Total number of units represented (Written in base ten notation)
1.	^{10}Box 3	3
2.	^{10}Box 8	_____
3.	^{43}Box 5	23
4.	^{100}Box 5	_____
5.	^{112}Box 4	_____
6.	^{112}Box 5	_____
7.	^{103}Box 5	_____
8.	^{211}Box 4	_____
9.	^{101}Box 8	_____
10.	^{999}Box 10	_____

Suppose you are given the amounts of wood below, before trading. Use the boxes of wood to make all possible trades. Record the COR-RECT FORM of the amount for each box. The first row has been completed for you.

Before trading	CORRECT FORM after all trades			
	Box 3	Box 4	Box 5	Box 10
1. Five units	12	11	10	5
2. Eight units	____	____	____	____
3. Six units	____	____	____	____
4. Four units	____	____	____	____
5. Two units	____	____	____	____
6. Twelve units	____	____	____	____
7. Eighteen units	____	____	____	____
8. Twenty-four units	____	____	____	____
9. Twenty-six units	____	____	____	____
10. Thirty units	____	____	____	____

NUMERATION SYSTEMS 5A

Upon successful completion of this lesson you should:

1. When given the CORRECT FORM in one base, be able to determine the CORRECT FORM for the same amount of wood (i.e. same total number of units) in another base.

2. When given the CORRECT FORM of two numbers in different bases, be able to determine which has the most units.

1. Suppose you take two LONGS and one UNIT from a box, other than Box 2. The CORRECT FORM for this wood is 21. How many units will you have if you take them from

Box 3? _____

Box 4? _____

Box 5? _____

2. Suppose you now take one FLAT, one LONG and two UNITS. The CORRECT FORM for this wood is 112. How many UNITS will you have if you take them from

Box 3? _____

Box 5? _____

Box 8? _____

3. Suppose you select wood from two boxes. If you record the CORRECT FORM for each box, can you tell which one has the most wood? Circle the box from which the greatest amount of wood was selected. Be sure to use the wood pieces whenever necessary.

   a. 103Box 4 103Box 5

   b. 1000Box 6 1000Box 3

   c. 111Box 4 111Box 5

4. Look at your answers to question #3. If the same numeral is used in different bases, in which base does it represent the greatest total number of units?_____

   Will this always be true? _____ If not, cite an example where it is not true.

5. For each of the following pairs, circle the one which has the greatest amount of wood. (i.e. greatest total number of units.)

   a. 143Box 5 132Box 6

   b. 333Box 4 112Box 7

   c. 110Box 4 222Box 3

   d. 221Box 3 101Box 5

   e. 103Box 4 103Box 6

   f. 1000Box 6 1000Box 3

6. Suppose you take two FLATS, two LONGS, and two UNITS (i.e. 222) from Box 4. Also suppose that one of your friends is asked to select the same amount of wood (NOT the same number of FLATS, LONGS and UNITS) from Box 6. If he selects correct pieces from Box 6, how should his answer be recorded?

   $$222\text{Box }4 = \underline{\quad} \text{ Box }6$$

7a. Now suppose you take three FLATS, two LONGS and one UNIT (321) from Box 4. Your friend is asked to select the same amount of wood from Box 3. If he selects the correct pieces how should be record his answer?

   $$321\text{Box }4 = \underline{\quad} \text{ Box }3$$

 b. If another friend selects the same amount of wood from Box 5, how should his answer be recorded?

   $$321\text{Box }4 = \underline{\quad} \text{ Box }5$$

8. Fill in the missing answers below:

   $$\underline{\quad} \text{ Box }4 = 120\text{Box }5 = \underline{\quad} \text{ Box }6$$
   $$\underline{\quad} \text{ Box }3 = \underline{\quad} \text{ Box }4 = 44\text{Box }5$$
   $$154\text{Box }8 = \underline{\quad} \text{ Box }9 = \underline{\quad} \text{ Box }10$$

NUMERATION SYSTEMS 6A

Upon successful completion of this lesson you should:

1. Be able to write the largest one digit number and its successor.
2. Be able to write the largest two digit number and its successor.
3. Be able to write the largest three digit number and its successor.
4. Be able to write the largest four digit number and its successor.

1. Remember our work with Box _3_ or Base _3._* Now circle the symbols which could appear in the CORRECT FORM if you are using Base _3._

    0   1 2 3 4 5 6 7 8 9

2. Write the CORRECT FORM of the largest number that can be expressed by a 1-digit numeral in Base _3._ _____

3. Take pieces from Base _3_ to represent the number recorded in question 2, then add one UNIT to your pieces. What is the CORRECT FORM? _____

4. Write the CORRECT FORM of the largest number that can be expressed by a 2-digit numeral in Base _3._ _____

5. Take pieces from Base _3_ to represent the number recorded in question 4, then add one UNIT to your pieces. What is the CORRECT FORM? _____

6. Write the CORRECT FORM of the largest number that can be expressed by a 3-digit numeral in Base _3._ _____

7. Take pieces from Base _3_ to represent the number recorded in question 6, then add one UNIT to your pieces. What is the CORRECT FORM? _____

---

* As noted earlier (page 154) the terms "box" and "base" are synonymous in these lessons. "Box" has been used almost exclusively in the first five lessons as it provides an excellent referent to the shoeboxes used to store the multibase pieces. Since "base" is the commonly accepted term to identify different grouping schemes, it will be used in this and remaining lessons. Such a change may seem abrupt, but we are sure you are now at the stage where these two terms can be mentally interchanged.

Select another Base. Using this base represent each of the following:

> Note: In Summary "A" write CORRECT FORM for answers to questions 8–11 respectively.
>
> In Summary "B" write CORRECT FORM after you have added one unit to each of the answers to questions 1–4 respectively.

	Summary "A"	Summary "B"
8. Largest number expressed as a one-digit numeral	_____	_____
9. Largest number expressed as a two digit numeral	_____	_____
10. Largest number expressed as a three digit numeral	_____	_____
11. Largest number expressed as a four digit numeral	_____	_____

12. Circle the symbols which could appear in the CORRECT FORM of a number with the base you are using.

    0   1   2   3   4   5   6   7   8   9

Using the ideas from above, complete the Table below.

	Base 3	Base 5	Base 6	Base 10
1. How many symbols are in...				
2. What is the largest number expressed as a one digit numeral in...				
3. What is the next whole number? (After adding one unit.)				
4. What is the largest number expressed as a two digit numeral in...				
5. What is the next whole number in...				
6. What is the largest number expressed as a three digit numeral in...				
7. What is the next whole number in...				

NUMERATION SYSTEMS 7A

Upon successful completion of this lesson you should:

1. Be able to add and subtract two, three and four digit numbers in particular bases.
2. Be able to determine the base given addition and/or subtraction problems that are solved correctly.

Solve the following addition and subtraction problems, using the blocks whenever you wish.

1.   14   Base 5
     (+)   3

2.   12   Base 5
     (−)   4

3.   24   Base 6
     (+) 12

4.   141   Base 5
     (−) 32

5.   121   Base 3
     (+) 22

6.   121   Base 4
     (+) 22

7.   214   Base 6
     (−) 25

8.   4122   Base 5
     (+) 323

    Check your answers to these exercises on the next page. If you have answered at least six correctly, go to the next questions. If you answered less than six of the problems correctly, work some of the practice problems from the supplementary problem sheet on page 175 and then come back to question #9.

9. Examine the following solution and then determine the base. Assume the problem is worked correctly.

         132
     (+) 13 Base_____
         200

10. Examine the following solution and then determine the base. Assume the problem is worked correctly.

       2314
     (−) 123 Base____
       2161

11. Look at this correct solution and then determine the base.

       13
       24
       21 Base_____
       42
      100

12. The following solution is incomplete. Determine the operation and the base and then *fill in the blanks* so that the solution is correct.

$$\begin{array}{r} 2\ \_\ 2\ 3 \\ 4\ \_\ 4 \\ \hline 3\ 1\ 2\ 2 \end{array}$$ Base _____

13. Examine this solution and then determine the base. Once again, assume it is worked correctly.

$$\begin{array}{r} 4\ 1\ 6 \\ (-)4\ 0\ 3 \\ \hline 1\ 3 \end{array}$$ Base _____

14. Bob says "any base will work" in the problem above. Once again Martha disagrees. Who is right and why? _____
_____

Answers to first eight problems on page 174:

1. ^{22}Base 5   2. ^{3}Base 5
3. ^{40}Base 6   4. ^{104}Base 5
5. ^{220}Base 3   6. ^{203}Base 4
7. ^{145}Base 6   8. ^{10000}Base 5

Supplementary Problem Sheet for Addition and Subtraction. Remember these are not for everyone. Answers to these are found at the bottom of the page.

1. $\begin{array}{r} 4\ 2\ 1 \\ (+)\ \ 3\ 3 \end{array}$ Base 5   2. $\begin{array}{r} 4\ 2\ 1 \\ (+)\ \ 3\ 3 \end{array}$ Base 6

3. $\begin{array}{r} 2\ 1\ 1\ 0 \\ (-)\ \ \ 2\ 1 \end{array}$ Base 3   4. $\begin{array}{r} 3\ 2\ 0\ 1 \\ (-)\ 1\ 0\ 2 \end{array}$ Base 6

5. $\begin{array}{r} 1\ 0\ 0\ 0 \\ (-)\ \ \ 1\ 1 \end{array}$ Base 3   6. $\begin{array}{r} 2\ 3\ 8\ 4 \\ (-)\ 7\ 2\ 5 \end{array}$ Base 10

7. $\begin{array}{r} 4\ 3\ 2\ 1 \\ (+)\ 3\ 2\ 2 \end{array}$ Base 5   8. $\begin{array}{r} 4\ 3\ 2\ 1 \\ (-)\ 3\ 2\ 2 \end{array}$ Base 5

9. $\begin{array}{r} 5\ 2\ 5 \\ (+)\ 3\ 5 \end{array}$ Base 6   10. $\begin{array}{r} 8\ 1\ 8 \\ (+)\ 7\ 5 \end{array}$ Base 9

Answers to Supplementary Problem Sheet

1. ^{1004}Base 5   2. ^{454}Base 6
3. ^{2012}Base 3   4. ^{3055}Base 6
5. ^{212}Base 3   6. ^{1659}Base 10
7. ^{10143}Base 5   8. ^{3444}Base 5
9. ^{1004}Base 6   10. ^{1004}Base 9

NUMERATION SYSTEMS 8A

Upon successful completion of this lesson you should:

1. Be able to determine the CORRECT FORM of the product in a particular base, given a multiplicand and a one digit multiplier.

2. Be able to determine the CORRECT FORM of a product in a particular base given a multiplicand and the base or power of its base as a multiplier.

1. Build a one-story house on top of the blueprint (Figure 1) using the wood represented by 231 in Box <u>4.</u> Now next to it build a two-story house using an identical blueprint. What is the CORRECT FORM of the amount of wood used to build the two-story house?

$$231 \times 2 = \underline{\hspace{2cm}} \text{ Base } \underline{4}$$

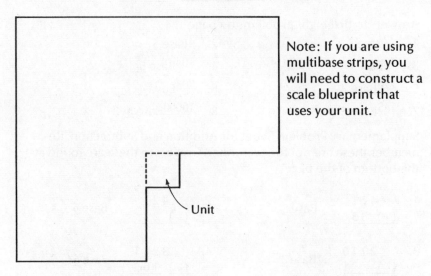

Note: If you are using multibase strips, you will need to construct a scale blueprint that uses your unit.

Unit

Figure 1

2. Now build a three-story house next to the two-story. Write the CORRECT FORM of the amount of wood used in this house.

$$231 \qquad \times 3 \qquad = \underline{\hspace{1.5cm}}$$
$$\text{Base } \underline{4} \quad \text{Base } \underline{4} \qquad \qquad \text{Base } \underline{4}$$

3. How much wood is in a four-story house with the same blueprint? (Remember to write the CORRECT FORM of both multiplier and product.)

$$231 \qquad \times \underline{\hspace{1.5cm}} \qquad = \underline{\hspace{1.5cm}}$$
$$\text{Base } \underline{4} \qquad \quad \text{Base } \underline{4} \qquad \quad \text{Base } \underline{4}$$

     Factor   Factor or   Product

                   multiplier*

---

* Either of the factors can be thought of as the multiplier. In this lesson the second factor is designated the multiplier.

4. This time take Box <u>3</u> and use the wood represented by 120 to build a one-story house on Figure 2. Build a two-story house next to it, using the same blueprint. How much wood did it take to build the second house? (Remember to write the CORRECT FORM)

120   × 2   =_____

Base <u>3</u>   Base <u>3</u>   Base <u>3</u>

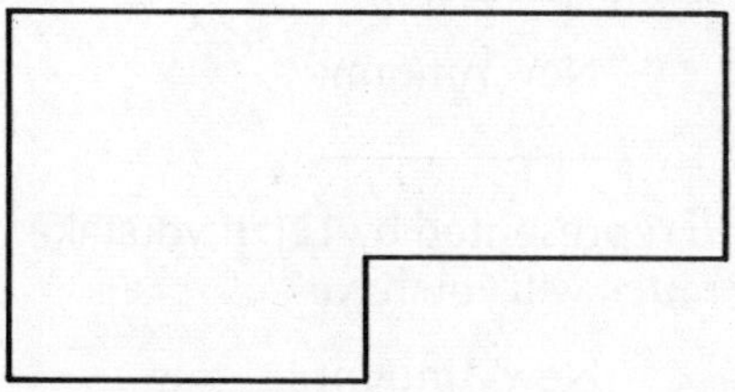

Figure 2

5. Now build a three-story house. How much wood did you use to build it? (Remember to write the CORRECT FORM of both the multiplier and product.)

120   ×_____   =_____

Base <u>3</u>   Base <u>3</u>   Base <u>3</u>

6. How much wood would you have to use to build a four-story house?_____

120   ×_____   =_____

Base <u>3</u>   Base <u>3</u>   Base <u>3</u>

7. If Mike has four longs and three units from Box <u>5</u> and you want twice as many as he has, how many of each piece would you take from Box <u>5</u>?

_____ LONGS _____ UNITS

8. Now write the CORRECT FORM for the pieces you selected in problem 7._____

9. Sue has the amount of wood represented by 121 from Box <u>5</u>. If you want three times as much wood, how many of each piece would you take from Box <u>5</u>?

_____

10. Write the CORRECT FORM of the amount of wood you selected in problem 9. _____

11. Suppose you take from Box 3 the wood represented by 221, i.e., the original amount. If a friend takes three times as much wood, what will be the CORRECT FORM of his amount i.e., new amount?

Original Amount   New Amount

_____   _____

12. Now take from Box <u>3</u> the wood represented by 11. Take out three times as much wood. Fill in these blanks.

Original Amount   New Amount

_____   _____

13. Using Box <u>4</u> take out the wood represented by 121. If you take out four times as much, how much will you have?

Original Amount   New Amount

_____   _____

14. Again use Box <u>4</u>. Take out four times as much wood as 32 represents from Box <u>4</u>. Fill in these blanks.

Original Amount   New Amount

_____   _____

15. From Box <u>6</u> take out the wood represented by 25. If you were to take out six times as much wood, how much would you have?

Original Amount   New Amount

_____   _____

16. Take from Box <u>6</u> six times the amount of wood represented by 100. How much wood do you have?

    Original Amount   New Amount

    _____   _____

17. There is no Box <u>9</u>. If you had such a box, and took the amount of wood represented by 518, what would be the representation of an amount nine times as much?

    Original Amount   New Amount

    _____   _____

18. Have you observed any patterns? If so briefly describe this pattern.

    _____

19. If you used the multiplication algorithm on each problem (11–17), what is the CORRECT FORM of the multiplier? _____

20. Now take from Box <u>5</u>, five times the amount of wood represented by 23 Base <u>5</u>. Write the CORRECT FORMS in this mathematical sentence which expresses the relationship.

    23 Base <u>5</u>   × _____   = _____
                          Base <u>5</u>   Base <u>5</u>.

21. Use the product from #20 and take five times this amount of wood. How much would you have?

    _____   × _____   = _____
    Base <u>5</u>   Base <u>5</u>   Base <u>5</u>.

22. Now suppose you take twenty five times the amount of wood represented by 23 Base <u>5</u>. How much would you have?

    23   × _____   = _____
    Base <u>5</u>   Base <u>5</u>   Base <u>5</u>.

23. How does the product in 21 compare with the product in #22?

    _____

NUMERATION SYSTEMS 9A

Upon successful completion of this lesson you should:

1. Be able to solve a division problem by partitioning (separating) a pile of wood into equivalent groups and recording the COR-RECT FORM.

2. Be able to determine the CORRECT FORM of a quotient in a particular base, given a dividend and a base or power of the base as a divisor.

Be sure to use the wood! Also remember to write all your answers in CORRECT FORM.

1. Suppose you have taken from Box 3 the amount of wood re-presented by this symbol: 22. Then a friend comes up and you give him exactly half of your wood. How much wood do you each have? (Write your answers in CORRECT FORM)

    Your wood   Your friend's wood
    _____   _____

2. Now suppose you have from Box 4 the amount of wood re-presented by 333. Two friends, Kelly and Scott, come up and you divide your wood equally among the three of you. How much wood do you each have?

    Your wood   Kelly's wood   Scott's wood
    _____   _____   _____

3. Suppose you have from Box 5 the amount of wood represented by this symbol: 432. Then your two friends come back and you have to divide the wood among the three of you again. How much wood do you each have this time?

    Your wood   Kelly's wood   Scott's wood
    _____   _____   _____

4. From Box 5 take the wood represented by 1414. Separate this evenly into three piles. Write the CORRECT FORM of the amount of wood in each pile.

    _____

5. Take 2140 from Box 6 and separate it evenly into four piles. How much will be in each of the piles?

    _____

Review of Division Algorithm

6. In solving problem number 1, could you use this kind of algorithm?

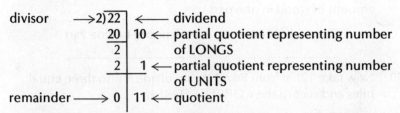

divisor ——→2)22 ←—— dividend
              20 10 ←— partial quotient representing number
               2   of LONGS
               2   1 ←— partial quotient representing number
                        of UNITS
remainder ——→ 0 11 ←— quotient

This quotient should, of course, equal your wood or your friend's wood.

7. Solve problem 2 again, using this algorithm. *Be sure to use the wood*, and then identify all parts below.

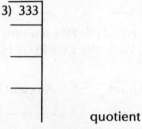

3) 333

                              quotient

8. Using this algorithm solve problems 3, 4, and 5 again.

3) 432   3) 1414   4) 2140

9. From Box <u>3</u> take the wood represented by 1210, (i.e., the original amount). Divide this into three equal parts. Write the CORRECT FORM for both the original amount of wood and the amount of wood in one part.

    Original Amount   Amount in One Part

    _____   _____

10. Now take 12120 from Box <u>3</u>. Again divide it into three equal piles and record the CORRECT FORM.

    Original Amount   Amount in One Pile

    _____   _____

11. If you have 2220 from Box <u>4</u> and separate it into four equal piles, how much will each pile have:

    Original Amount   Amount in One Pile

    _____   _____

12. Separate into four equal piles the amount of wood represented by 130 from Box <u>4</u>. Write the CORRECT FORM of the wood in each pile.

    Original Amount   Amount in One Pile

    _____   _____

13. Make six equal piles out of the wood represented by 2340 from Box <u>6</u>. How much wood do you have in each pile?

    Original Amount   Amount in One Pile

    _____   _____

14. There is no Box 9. However if you could take 16820 from a Box 9 and divided it into nine equal piles, could you guess how much wood would be in each pile?

    Original Amount   Amount in One Pile

    _____   _____

15. Have you observed any patterns? If so briefly describe this pattern.

    _____

16. If you used the division algorithm on each problem (9-14), what was the CORRECT FORM of the divisor?_____

17. Suppose you wanted to divide $23400_{Base\ 5}$ equally among twenty-five students:

    What would be the CORRECT FORM of the divisor?_____

    How much would each student receive?_____

18. Suppose you wanted to divide $23000_{Base\ 4}$ equally among sixteen students:

    What would be the CORRECT FORM of the divisor?_____

    How much would each student receive?_____

19. Suppose you wanted to divide $23000_{Base\ 4}$ equally among sixty-four students:

    What would be the CORRECT FORM of the divisor?_____

    How much would each student receive?_____

Whew! Nine sequential activities for studying numeration systems have been presented. Hopefully you have been actively involved in these lessons and have achieved the stated objectives for the respective lessons. In fact, at this point you may wish to refer to the specific behavioral objectives to help tie together loose ends and provide closure to these lessons. Test yourself. How did you do? We're confident that if you honestly gave these activities the "old college try" you were able to satisfactorily complete them.

If you have grasped all of the fundamental ideas identified by the lesson objectives, then you may skip the following set of parallel lessons and move directly to the post activities. On the other hand if you have questions about some of the ideas, or would like to explore them in a different physical embodiment, then you should examine the next series of lessons entitled "Numeration Systems Via Lightbulbs."

The Flowchart (page 151) shows that this is a parallel set of activities with each lesson having a respective counterpart in the Multibase Blocks lessons. The first lesson establishes the new terminology that is used in remaining lessons. Once you complete the initial lesson you can work the lessons consecutively or you may select only those lessons that reflect specific ideas that you would like to explore in another setting. For example, if you wanted to examine "place value" in this series, you could move directly to the Lighthouse Game (NS-3B).

NUMERATION SYSTEMS VIA LIGHTBULBS

These lessons have been developed to provide an example of multiple embodiment or perceptual variability for the topic numeration systems. They are designed as parallel lessons although it is assumed that the reader has completed the companion lessons with the multibase blocks. The format for the lightbulbs and the multibase blocks are similar.

Once again these lessons are preceeded by specific objectives. These objectives focus attention on the central ideas developed in the lessons. You should review these objectives at the end of each lesson and check to see that each has been achieved.

OVERVIEW OF LESSON OBJECTIVES

Upon successful completion of the following numerations systems you should:

NS-1B
1. Be able to state relationships among Yellow, Blue, Green, Red, etc. lightbulbs for a given house.
2. Be able to make fair trades using lightbulbs within a given house.

NS-2B
1. When given a collection of lightbulbs and a particular house, be able to make all necessary trades, write the result in CORRECT FORM and read it properly.
2. When working with a given house be able to determine which symbols could appear in CORRECT FORM.

NS-3B
1. When given CORRECT FORMS of different numbers in a particular base, be able to determine by inspection their ordinal value, i.e., their order relationship.
2. Be able to write the largest three digit whole number in a given base.
3. Be able to analyze the Lighthouse Game to determine the eventual outcome.

NS-4B
1. Be able to determine by inspection the number of different colored lightbulbs when given the CORRECT FORM.
2. When given a total number of Yellow bulbs, be able to determine the CORRECT FORM for a particular base.
3. When given the CORRECT FORM and particular house, be able to determine the total number of yellow bulbs represented.

NS-5B

1. When given the CORRECT FORM in one base, be able to determine the CORRECT FORM for the same amount of light (i.e. same total number of yellow bulbs) in another base.

2. When given the CORRECT FORM of two numbers in different bases, be able to determine which has the most yellow bulbs.

NS-6B

1. Be able to write the largest one, two, three and four digit numbers and their respective successor.

2. Be able to write one, two, three and four digit numbers and their respective successor.

NS-7B

1. Be able to add and subtract two, three and four digit numbers in particular bases.

2. Be able to determine the base given addition and/or subtraction problems that are solved correctly.

3. Be able to analyze addition and subtraction problems and determine relevant data that are missing.

NS–8B

1. Be able to determine the CORRECT FORM of the product in a particular base, given a multiplicand and a one digit multiplier.

2. Be able to determine the CORRECT FORM of a product in a particular base given a multiplicand and the base or power of its base as a multiplier.

NS–9B

1. Be able to solve a division problem by partitioning (separating) the lightbulbs into equivalent groups and recording the CORRECT FORM.

2. Be able to determine the quotient in a particular base, given a dividend and a base or power of the base as a divisor.

Teachers using light house cards

NUMERATION SYSTEMS 1B

These are lightbulbs:

Pink   Orange   Red   Green   Blue   Yellow

⊙   ⊙   ⊙   ⊙   ⊙   ⊙

This is a house card:*

Orange Room   Red Room   Green Room   Blue Room   Yellow Room

Front of House card

These are sockets for the lightbulbs in House 4

This house card has rooms arranged in a row. Only yellow lightbulbs can be put into the yellow room; only blue lightbulbs can be put into the blue room; only green lightbulbs can be put into the green room, etc.

Back of House card

House Number _____

_____ Yellow lightbulbs may be traded for _____ Blue.

_____ Blue lightbulbs may be traded for _____ Green.

_____ Green lightbulbs may be traded for _____ Red.

_____ Red lightbulbs may be traded for _____ Orange.

---

* Pupils can make their own house card. Here are some suggestions:
1. Poster board or construction paper cut in $1\frac{1}{2}''$ by $7\frac{1}{2}''$ strips make fine housecards. The borders on the front of the housecard can be made any color. However once a definite color arrangement has been established (such as orange to yellow in these lessons) it should be maintained.
2. The number of light sockets in each room may be drawn on the front of the house card. Some people find these sockets confusing and prefer to keep them in mind rather than sketch them on each house card. In either case the number of sockets in each room is always one less than the house number. In House 4 for example, each of the rooms would have three sockets; so the greatest number of light bulbs in any House 4 room is three.

Here are the backs of house cards that will be used in these lessons.

House Number __2__

__2__ Yellow lightbulbs may be traded for __1__ Blue.

__2__ Blue lightbulbs may be traded for __1__ Green.

__2__ Green lightbulbs may be traded for __1__ Red.

__2__ Red lightbulbs may be traded for __1__ Orange.

House Number __3__

__3__ Yellow lightbulbs may be traded for __1__ Blue.

__3__ Blue lightbulbs may be traded for __1__ Green.

__3__ Green lightbulbs may be traded for __1__ Red.

__3__ Red lightbulbs may be traded for __1__ Orange.

House Number __4__

__4__ Yellow lightbulbs may be traded for __1__ Blue.

__4__ Blue lightbulbs may be traded for __1__ Green.

__4__ Green lightbulbs may be traded for __1__ Red.

__4__ Red lightbulbs may be traded for __1__ Orange.

House Number __5__

__5__ Yellow lightbulbs may be traded for __1__ Blue.

__5__ Blue lightbulbs may be traded for __1__ Green.

__5__ Green lightbulbs may be traded for __1__ Red.

__5__ Red lightbulbs may be traded for __1__ Orange.

3. Trading rules for each house card should be clearly printed on the back.

4. Lightbulbs can be made with a paper punch. You can punch holes in appropriately colored construction paper or punch holes in white paper and color it. In either case the "colored holes" make excellent light bulbs. The number of light bulbs depends on the activity. Generally more yellow bulbs (approximately 20) are needed than any of the others (approximately 15 of each remaining color.)

House Number ___6___

___6___ Yellow lightbulbs may be traded for ___1___ Blue.

___6___ Blue lightbulbs may be traded for ___1___ Green.

___6___ Green lightbulbs may be traded for ___1___ Red.

___6___ Red lightbulbs may be traded for ___1___ Orange.

House Number ___10___

___10___ Yellow lightbulbs may be traded for ___1___ Blue.

___10___ Blue lightbulbs may be traded for ___1___ Green.

___10___ Green lightbulbs may be traded for ___1___ Red.

___10___ Red lightbulbs may be traded for ___1___ Orange.

The front of each house card looks the same unless you insert the light sockets. However on the back of each house card is the number of the house, and information telling you how many yellow lightbulbs can be traded for a blue lightbulb, how many blue lightbulbs can be traded for a green lightbulb, and how many green lightbulbs can be traded for a red lightbulb, etc.

Take out a house card and some lightbulbs, and then answer the following questions:

1. The house card number is_____

2. How many sockets are there in each room?_____

3. How many yellow lightbulbs does it take to fill the yellow room? _____

4. How many blue lightbulbs does it take to fill the blue room?_____

5. How many green lightbulbs does it take to fill the green room? _____

6. How many red lightbulbs does it take to fill the red room?_____

On the house card it tells you how many yellow lightbulbs can be traded for a blue lightbulb, and how many blue lightbulbs can be traded for a green lightbulb, and so on. Using this information answer the following questions.

1. How many yellows trade for a blue?_____

2. How many yellows trade for a green?_____

3. How many yellows trade for a red?_____

4. How many yellows trade for an orange?_____

5. How many blues trade for a green?_____

6. How many blues trade for a red?_____

7. How many greens trade for a red?_____

8. How many greens trade for an orange?_____

9. How many reds trade for an orange?_____

NUMERATION SYSTEMS 2B

Take 5 Green bulbs, 4 Blue bulbs and 2 Yellow bulbs.

GREEN	BLUE	YELLOW	
5	4	2	Before trade

Use House card <u>4</u> and make all necessary trades so your bulbs should look like this:

RED	GREEN	BLUE	YELLOW	
1	2	0	2	After All Trades (House card <u>4</u>)

Write this as 1202 which is called the CORRECT FORM and read as "'one', 'two', 'zero', 'two' House <u>4</u>" or "'one', 'two', 'zero', 'two' Base <u>4</u>".

Suppose that instead of House card <u>4</u> you take the same bulbs (i.e. 5 Green, 4 Blue and 2 Yellow) and use House card <u>5.</u>

1. Make all necessary trades and record your results below:

RED	GREEN	BLUE	YELLOW	
				After All Trades (House card <u>5</u>)

2. The CORRECT FORM is_____ and should be read as

_____.

Now, take out a house card that has three sockets in each room, *not* House card 3. Suppose you received a shipment of 6 Red bulbs, 3 Green bulbs, 2 Blue bulbs, and 4 Yellow bulbs. Now *trade so that you can obtain the same candlepower using the fewest number of lightbulbs.*

1. When you have finished trading, you should have:

ORANGE	RED	GREEN	BLUE	YELLOW
1	*2*	*3*	*3*	*0*

2. The CORRECT FORM is_____ and should be READ as
    _____.

3. What does the "0" in the CORRECT FORM tell us?_____
    _____

Answer these questions using the same House card.

1. What is the greatest number of bulbs you can have in each room of this house?_____

2. What is the least number of bulbs you can have in each room of this house? (Think about this before you put down an answer.)
    _____

3. What is the greatest digit you can use in CORRECT FORM with this house?_____

4. What base do you suppose you are working with?_____

Select any House card and record it here: House card_____

1. Could a "5" appear in the CORRECT FORM?_____

2. Circle the symbols that *could* appear in the CORRECT FORM.
    0, 1, 2, 3, 4, 5, 6, 7, 8, 9.

NUMERATION SYSTEMS 3B

LIGHT HOUSE GAME

Directions:

1. Play this game in small groups. (Approximately 2–4)

2. Use either a die or spinner as appropriate for a given Lighthouse Map. (See diagram below)

3. All players start from the same point. The roads they travel are determined by the die or spinner. After each spin or roll a player moves along the road to the sign corresponding to the numeral determined by the spin or roll. (This can be done by either moving a marker or actually tracing moves on the map.) The player remains at this sign until his next roll.

4. A player spins (or rolls) and then moves. A player must move from one sign directly to another and cannot at anytime move backward. The next player spins and moves. Play continues until each player has had three turns.

5. After each person has had three turns, the winner will be the player closest to the Lighthouse.

6. Be sure to record the roads you travel on the score sheet.

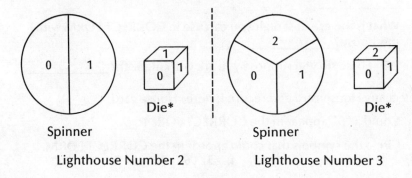

Spinner   Die*   Spinner   Die*

Lighthouse Number 2   **Lighthouse Number 3**

---

* Only three faces are shown on these dice. The hidden faces would also have to be labeled.

LIGHTHOUSE MAP

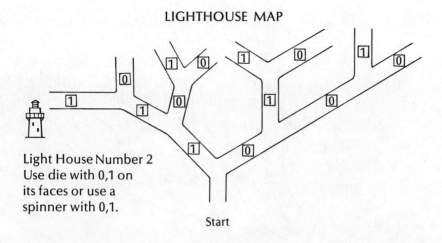

Light House Number 2
Use die with 0,1 on
its faces or use a
spinner with 0,1.

Start

Play this game several times and record the results of two of your
games below.

NAME	1ST turn	2ND turn	3RD turn

NAME	1ST turn	2ND turn	3RD turn

LIGHTHOUSE MAP

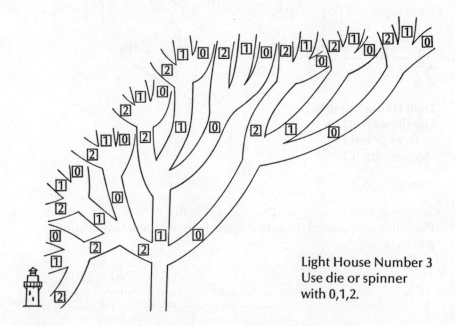

Light House Number 3
Use die or spinner
with 0,1,2.

Play this game several times and record the results of two of your games below.

NAME	1ST Turn	2ND Turn	3RD Turn

NAME	1ST Turn	2ND Turn	3RD Turn

After you have played the game several times consider these questions:

1. When could you predict who would end up closest to the Lighthouse?

    _____

2. When could you predict who would end up farthest from the Lighthouse?

    _____

3. What series of throws (or spins) would put you closest to Lighthouse Number 3?

    _____

4. How does your answer to the above question compare to the largest three digit whole number in base three?

    _____

Consider the following summary for Lighthouse Number 6.

NAME	1ST Turn	2ND Turn	3RD Turn
Scott	2	4	5
Cheryl	3	1	2
Kelly	4	3	1
Timmy	4	1	5

1. Who was closest to Lighthouse 6? _____ .
2. Who was farthest from Lighthouse 6? _____ .

NUMERATION SYSTEMS 4B

If you pick up a bulb, we don't know how much light it will provide until we learn its color. If you choose two Green bulbs we would report it as 200.

1. How many of each color bulb would these CORRECT FORMS represent?

ORANGE	RED	GREEN	BLUE	YELLOW	CORRECT FORM
____	____	____	____	____	4
____	____	____	____	____	103
____	____	____	____	____	4010
____	____	____	____	____	410

2. Determine how many Yellow bulbs are represented by each of the following:

CORRECT FORM	TOTAL NUMBER OF YELLOW BULBS REPRESENTED
a. ^{21}House $\underline{3}$	_____
b. ^{21}House $\underline{4}$	_____
c. ^{21}House $\underline{5}$	_____
d. ^{103}House $\underline{6}$	_____
e. ^{103}House $\underline{5}$	_____
f. ^{103}House $\underline{4}$	_____
g. ^{876}House $\underline{10}$	_____

3. Take our House Card $\underline{3}$, House Card $\underline{6}$, and House Card $\underline{10}$. Suppose you received shipments of different amounts of Yellow lightbulbs listed below. Write the CORRECT FORM after trading, for each of these houses.

SHIPMENT	House $\underline{3}$	House $\underline{6}$	House $\underline{10}$
3 Yellow bulbs			
6 Yellow bulbs			
10 Yellow bulbs			
15 Yellow bulbs			
27 Yellow bulbs			
38 Yellow bulbs			

1. Take out House 5 for yourself, and let your friend work with House 7 . Now suppose a shipment of twenty-one Yellow light-bulbs came to you and an identical shipment came to your friend. You can't light your house correctly with twenty-one Yellow lightbulbs, and neither can your friend. Therefore make whatever trades are necessary and write the CORRECT FORM for both your house and your friend's house.

<table>
<tr><td>YOUR HOUSE
(House 5)</td><td>FRIEND'S HOUSE
(House 7)</td></tr>
<tr><td>_____</td><td>_____</td></tr>
</table>

2. Can the lights in House 5 and House 7 each be traded back for twenty-one Yellow lightbulbs?_____

3. Does ^{41}House 5 = ^{30}House 7?_____

4. Let's suppose you have other friends with different houses. If you were to compare a number in one house to a number in another house, as shown below, who would have the greatest amount of light? (i.e. whose lights could be traded back for the most number of Yellow lightbulbs.) Circle the largest.

   a. ^{10}House 3 ^{10}House 5

   b. ^{131}House 5 ^{131}House 4

   c. ^{122}House 9 ^{122}House 10

5. If the same CORRECT FORM appears in different houses, can you determine in which of these houses it represents the greatest amount of light? _____

   If so, what would be your clue? _____

6. Suppose you have House 8 , and the rooms are lit with 132. If your friend has House 6 , how would he light his house using the same amount of light that you have? In other words, so that you both have the same total number of Yellow lightbulbs.

   ^{132}House 8 = _____ House 6

7. Now suppose you had 32 in House 10, and another friend had House 5 and wanted to light his house with the same amount of light you have. How would he light his house?

   ^{32}House 10 = _____ House 5

NUMERATION SYSTEMS 6B

1. What is the CORRECT FORM of the largest number of Yellow bulbs that can be represented by a 1-digit numeral in House 5.? _____

2. If one Yellow bulb is added to this group, how would the CORRECT FORM of the result be represented? _____

3. What is the CORRECT FORM of the largest number of Blue and Yellow bulbs that can be represented by a 2-digit numeral in House 5? _____

4. If one Yellow bulb is added to this group, how would the CORRECT FORM of the result be represented? _____

5. If one Green bulb is added to this group, (exercise 3), how would the CORRECT FORM of the result be represented? _____

6. Suppose you have the CORRECT FORM of light bulbs reported in the left column. What is the result when:

CORRECT FORM	One Yellow bulb is added?	One Blue bulb is added?	One Green bulb is added?
^{2}House 3	_____	_____	_____
^{55}House 6	_____	_____	_____
^{333}House 4	_____	_____	_____

7. Now take House 7 and light it with 6666. If you decided to add one Yellow bulb what would be the CORRECT FORM of the result? _____

8. Suppose a friend has been watching you. He added one Yellow bulb to his lights and got 10000 in House 10. What do you think he had before he added the one Yellow Bulb? _____

9. In whose house (7 or 10) would the 10000 mean the most light? (i.e. worth the most Yellow bulbs)? _____

10. Suppose Bob adds a Green bulb to his lights and got 10002 in House 5. What was the CORRECT FORM of his lights before he added the Green bulb? _____

Solve the following addition and subtraction problems using the House cards whenever you wish. Each is written in CORRECT FORM.

1.   23   House <u>5</u>
    (+) 4

2.   14   House <u>6</u>
    (−) 5

3.   21   House <u>5</u>
    (+)14

4.   123   House <u>4</u>
    (−) 32

5.   123   House <u>6</u>
    (+) 33

6.   142   House <u>8</u>
    (+) 31

7.   1010   House <u>3</u>
    (−) 202

8.   2143   House <u>5</u>
    (+) 302

Check your answers to these problems below. If you answered at least 6 correct go to the questions below. Otherwise work some of the practice problems on page 202 and then return to the questions below.

9. Examine the following solution and then determine the House. Assume it is worked properly!

$$\begin{array}{r} 2341 \\ (+)\,1543 \\ \hline 4324 \end{array}$$ House _____

10. Bob worked the following problem with House <u>5</u>.

$$\begin{array}{r} 331 \\ (+)\,312 \\ \hline 643 \end{array}$$

Can you find something wrong with Bob's solution? _____
Explain _____

11. Determine the missing subtrahend in this problem:

$$\begin{array}{r} 2341 \\ (-)\,\text{-- -- -- --} \\ \hline 132 \end{array}$$ House <u>5</u>

12. Here are parts of a problem that have been worked correctly. Can you find the House and operation being used as well as the missing digit?

$$\begin{array}{r} 312 \\ (\ \ )\,1\_4 \\ \hline 134 \end{array}$$ House _____

Answers to first eight problems on previous page.

1. ^{32}House 5
2. ^{5}House 6
3. ^{40}House 5
4. ^{31}House 4
5. ^{200}House 6
6. ^{173}House 8
7. ^{101}House 4
8. ^{3000}House 5

Additional Problems for Addition and Subtraction. Remember, these are not for everyone. Answers to these are shown below.

1. $\quad$ 31 $\quad\quad$ House 7
    $\underline{(-) 1\,2}$

2. $\quad$ 412 $\quad\quad$ House 6
    $\underline{(+)\quad 5\,3}$

3. $\quad$ 512 $\quad$ House 6
    $\underline{(-)\quad 4\,3}$

4. $\quad$ 1201 $\quad\quad$ House 5
    $\underline{(-)\quad 3\,4\,2}$

5. $\quad$ 1021 $\quad$ House 5
    $\underline{(-)\quad 3\,4\,2}$

6. $\quad$ 1212 $\quad\quad$ House 3
    $\underline{(+)\quad 1\,0\,2}$

7. $\quad$ 212 $\quad$ House 4
    $\underline{(+)1\,3\,1}$

8. $\quad$ 1001 $\quad\quad$ House 3
    $\underline{(-)\quad 2\,0\,2}$

Answers to additional problems.

1. ^{16}House 7
2. ^{505}House 6
3. ^{425}House 6
4. ^{304}House 5
5. ^{124}House 5
6. ^{2021}House 3
7. ^{1003}House 4
8. ^{22}House 3

NUMERATION SYSTEMS 8B

Be sure to use the House cards! Also remember to write all of your
answers in CORRECT FORM.

1. Suppose you have one Blue and two Yellow
    lightbulbs from House 4. (i.e. 12 House 4)
    If Kelly wants *twice* as much light (i.e. twice as
    many of each bulbs) as you now have, what
    would be the CORRECT FORM of the bulbs
    she should select from House 4?   _____

2. If Tim wants *three times* as much light as you
    have, what would be the CORRECT FORM
    of the bulbs he should select from House 4?   _____

3. If Scott wants four times as much light as
    you have, what would be the CORRECT
    FORM of the bulbs he should select from
    House 4?   _____

4. This time take two Green and One Yellow
    lightbulb from House 5. (201 House 5). If
    Joan is to have *twice* as much light as you
    have, what would be the CORRECT FORM
    of the bulbs she should select from House
    5?   _____

5. If Sally wants *four times* as much light as you
    have, what would be the CORRECT FORM
    of the bulbs she should select from House
    5?   _____

6. If Martha wants *five times* as much light as
    you have, what would be the CORRECT
    FORM of the bulbs she should select from
    House 5?   _____

7. Suppose you take bulbs represented by 21 from House 3. If a
    friend takes *three times as much* light from House 3 what would
    be the CORRECT FORM of his amount?

    CORRECT FORM of   CORRECT FORM of
    Your Lightbulbs   Your Friend's Lightbulbs

    _____   _____

8. Suppose you take bulbs represented by 312 House 4. If a friend
    wants *four times as much* light from House 4 what would be the
    CORRECT FORM of his amount?

    CORRECT FORM of   CORRECT FORM of
    Your Lightbulbs   Your Friend's Lightbulbs

    _____   _____

203

9. Now take 213 from House 5. If a friend takes *five times as much* light from House 5 what would be the CORRECT FORM of his amount?

   CORRECT FORM of
   Your Lightbulbs

   _____

   CORRECT FORM of
   Your Friend's Lightbulbs

   _____

10. This time take 124 from House 6. If a friend takes *six times as much* light from House 6, what should be the CORRECT FORM of his amount?

    CORRECT FORM of
    Your Lightbulbs

    _____

    CORRECT FORM of
    Your Friend's Lightbulbs

    _____

11. Suppose you take 215 from House 10. If a friend takes *ten times as much* light from the House 10, what should be the CORRECT FORM of his amount?

    CORRECT FORM of
    Your Lightbulbs

    _____

    CORRECT FORM of
    Your Friend's Lightbulbs

    _____

12. If you observe any pattern, briefly describe it._____
    _____

13 Fill in the missing blanks:

    a. 210 House 3 is _____ times as much light as 21 House 3.

    b. 210 House 4 is _____ times as much light as 21 House 4.

    c. 210 House 10 is _____ times as much light as 21 House 10.

    d. 210 House 12 is_____ times as much light as 21 House 12.

14. See if you can also fill in these missing blanks:

    a. 42000 House 5 is _____ times as much light as 4200 House 5.

    b. 4200 House 5 is _____ times as much light as 420 House 5.

    c. 420 House 5 is _____ times as much light as 42 House 5.

15. Use the information from question 14 to determine how 42000 House 5 compares to 42 House 5.

Be sure to use the House cards! Also remember to write all of your answers in CORRECT FORM.

1. Suppose you have lightbulbs represented by ^{44}House $\underline{5.}$ A friend comes and you give him exactly half your lightbulbs. How many lightbulbs do each of you have?

CORRECT FORM of Your Lightbulbs	CORRECT FORM of Your Friend's Lightbulbs
_____	_____

2. Suppose that you again have lightbulbs represented by ^{44}House $\underline{5.}$
    Two friends, Kelly and Scott, come and you divide your lightbulbs into three equal groups. How many lightbulbs do each of your friends have?

CORRECT FORM of Your Lightbulbs	CORRECT FORM of Kelly's Lightbulbs	CORRECT FORM of Scott's Lightbulbs
_____	_____	_____

3. Once again let's have lightbulbs represented by ^{44}House $\underline{5.}$ Divide these lightbulbs into four equal groups and determine the CORRECT FORM of each group.

CORRECT FORM of Original Lightbulbs	CORRECT FORM of Each group (Four groups)
^{44}House $\underline{5}$	_____

4. For the fourth and final time let's suppose that you have lightbulbs represented by ^{44}House $\underline{5.}$ Divide the lightbulbs into six equal groups and determine the CORRECT FORM of each group.

CORRECT FORM of Original Lightbulbs	CORRECT FORM of Each Group (Six groups)
44	_____

5. From House $\underline{2}$ take lightbulbs represented by 110. Divide this into two equal groups. Write the CORRECT FORM for both the original amount of lightbulbs and the lightbulbs in each group.

ORIGINAL AMOUNT	EACH GROUP
_____	_____

6. Now take 420 from House $\underline{5.}$ Divide this into five equal groups and record the CORRECT FORM for the following:

ORIGINAL AMOUNT	EACH GROUP
_____	_____

7. If you take 1020 from House 3 and separate it into three equal groups, how many lightbulbs would each group have?

   ORIGINAL AMOUNT   EACH GROUP

   _____   _____

8. Separate into four equal groups the lightbulbs represented by 130 House 4.

   ORIGINAL AMOUNT   EACH GROUP

   _____   _____

9. Make six equal groups from the lightbulbs represented by 2540 House 6.

   ORIGINAL AMOUNT   EACH GROUP

   _____   _____

10. If you take 1270 from House card 10 and divide the lightbulbs into ten equal groups, how many lightbulbs would be in each group?

    ORIGINAL AMOUNT   EACH GROUP

    _____   _____

11. Have you observed any patterns?   If so briefly describe.

    _____

12. Suppose you wanted to divide ^{23400}House 5 among five students. Each student would get_____ House 5.

13. Suppose you take your answer in the previous problem (exercise 12) and divide each student's lightbulbs into five groups. Each group would have _____ House 5.

14. Suppose you wanted to divide ^{23400}House 5 into twenty five groups. Each group would have _____ House 5.

15. How does your answer to #14 compare to your answer to #13?

    _____

16. Suppose you wanted to divide ^{2100}House 3 into nine equal groups. Each group would have_____ House 3.

17. Suppose you wanted to divide ^{31200}House 4 into sixteen equal groups. Each group would have_____ House 4.

18. Suppose you wanted to divide ^{100000}House 2 into sixteen equal groups. Each group would have_____ House 2.

SOME POST LESSON PARALLEL ACTIVITIES*

There are many activities we could now consider. Among other things we could develop additional lessons that explore different numeration systems in greater depth (for example, work with fractions and exponents); we could provide another sequence of parallel lessons; we could construct specific worksheets (to increase skill in addition and subtraction, for example); or we could construct a test that reflects the lesson objectives. Each of these activities would probably be beneficial for some pupils. However, it is the teacher who must decide what kinds of activities are needed; when they should be used; how they should be developed; and which pupils should be involved. Although only the teacher can answer these questions, we are presenting several Post Lesson Activities for consideration.

POST ACTIVITY NUMBER 1

Another embodiment for numeration systems:

Suppose you travel to a foreign country that has the following currency:

SYMBOL	U. S. CURRENCY
A	$ .01
E	.05
I	.25
O	1.25
U	6.25

You want to trade your money for that of the new country so that you have the least amount of coins to carry.

1. For what would you trade the following amounts:

	U	O	I	E	A	Representation
.48	_	_	1	4	3	143
$3.56	_	_	_	_	_	_____
$5.36	_	_	_	_	_	_____
$13.50	_	_	_	_	_	_____

2. When you are ready to return home if you have an amount represented by 12034, how much U. S. money can you exchange it for? _____

3. What base is this foreign country using? _____

---

* Since evaluation of pupil progress permeates the entire teaching-learning experience, it will not be considered here as a separate entity. Chapter 10 explains our viewpoint on evaluation in the laboratory setting.

POST ACTIVITY NUMBER 2

Another embodiment for numeration systems.

These are weights·*

This is a SMALL.

This is a MEDIUM.

This is a LARGE.

This is an EXTRA-LARGE.

Use the scales to answer these questions:

1. How many SMALLS make a MEDIUM? _____

2. How many MEDIUMS make a LARGE? _____

3. How many LARGES make an EXTRA-LARGE? _____

4. How many SMALLS make a MEDIUM? _____

5. How many SMALLS make a LARGE? _____

6. How many SMALLS make an EXTRA-LARGE? _____

7. What base are you operating with? _____

---

*   Small plastic containers (found in any housewares department) were
     used in this lesson. At least four containers of each type (SMALL,
     MEDIUM, etc.) are needed. Sand was placed in these containers accord-
     ing to two rules:
     1. Each container of a given type weighs the same.
     2. A definite ratio** is established and maintained in moving from
         one size container to another. For example, if the SMALL weighs
         $\frac{1}{3}$ of the MEDIUM, then the MEDIUM would weigh $\frac{1}{3}$ of the
         LARGE, and the LARGE would weigh $\frac{1}{3}$ of the EXTRA-LARGE.
     A small scales or balance should be provided so that pupils can deter-
     mine the weight relationships that exist among the various containers.

** Mathematical variability can be provided by changing the weight ratios among
     the various containers.

PART A

Numeration systems with bases greater than ten.

1. Recall your past work with the bases. The digits needed to represent all numbers in base *three* are _____.

2. What are the digits used in base *five*? _____

3. In general, how many different digits are needed to represent all numbers in, say, base "*b*"? (Remember, 0 is a digit!)
   _____

4. For base *ten* you use all of the digits which you are familiar with. What would you do if you wanted to use a base larger than ten?
   _____

5. Say you wanted to use base *eleven*. How many digits would be needed? _____

6. Although there is no universal consensus, people usually agree to call the tenth digit (excluding zero) for base *eleven*, T. What do you think "T" represents? _____

7. Complete the list of digits for base *eleven*: 0, 1, 2, _____

8. Representations in base eleven may look strange to you at first. For example, if you have ten things, the correct form would be T; for sixteen it is <u>15</u>; for twenty-one it is <u>1 T</u>; and for twenty-two it is <u>20</u>.

9. Translate from the CORRECT FORM in base ten below to the equivalent CORRECT FORM in base eleven.

	CORRECT FORM BASE *TEN*	CORRECT FORM BASE *ELEVEN*
Fifteen	_____	_____
Twenty-seven	_____	_____
Thirty	_____	_____
Thirty-two	_____	_____
Thirty-three	_____	_____

10. Write the amount of wood (in base ten) represented by each of the following:

   35Base <u>eleven</u>   _____

   10TBase <u>eleven</u>   _____

   TTBase <u>eleven</u>   _____

11. What is the next largest base after Base <u>eleven?</u>   _____

12. How many digits will you need for this base? _____

13. What symbol would you suggest using for the last digit, recalling the representation in Base <u>eleven</u>? (i.e. T for ten) _____

PART B

Suppose you have eggs, egg cartons and shipping boxes that each hold one dozen egg cartons.

Egg Carton

Top View   Side View

Shipping Box

1. How many eggs are in each egg carton?   _____

2. How many cartons are in a shipping box?   _____

3. How many eggs are in a shipping box?   _____

Let's agree to represent this as we have done before. That is with the smallest piece on the right and move back to the left. Then we will have BOXES, CARTONS, EGGS.

Suppose a grocer places an order for 120: i.e.,

$$\underset{\text{BOXES}}{\underline{1}} \qquad \underset{\text{CARTONS}}{\underline{2}} \qquad \underset{\text{EGGS}}{\underline{0}}$$

How many eggs would he actually be ordering?

That is, 120Base <u>twelve</u> = ——— Base <u>ten.</u>

4. For each of the orders below, how many eggs are actually wanted?

Base <u>twelve</u>	Number of eggs
209	_____
1TE	_____
T 0 E	_____
10TE	_____

5. How would the orders read for each of the following?

Eggs	Representation in Base <u>twelve</u>
One-hundred thirty-six	_____
One-hundred fifty-four	_____
Fifty-seven	_____

6. Base <u>twelve is called</u> the duo-decimal system. Can you think of other physical things that work in this system?

Teachers experiencing another embodiment of numerical systems

POST ACTIVITY NUMBER 4

Numeration systems with different symbols.

Here are the first twelve counting symbols in a different numeration system. (i.e. they correspond from left to right to 1, 2, 3, 4, 5, . . . 12 in our Hindu-Arabic system)

| S E H S XI II SI EI HI SI XS

(Mirror line)

Use this information as well as what you have previously learned about numeration systems to answer the following questions:

1. How many different symbols are used?_____

2. Do you think this system has place value?_____
    How is the place value used here different from the place value
    we have been using? _____

3. What is the base of this numeration system? _____

4. The next two counting numbers to follow X S are_____ and
    _____ respectively.

5. The largest two digit number is represented by_____.

6. The largest four digit number is represented by_____.

7. The counting number immediately preceding X H is_____.

8. Which of the following represents the largest number?

IXS ISX XIS SIX SXI XSI

    Why?_____

9. Write the smallest number that you can using the symbols X ,
    E and S once and only once._____

Place the edge of a mirror on the Mirror Line. Look into the mirror and use the image to answer the following questions:

10. How many of the symbols look familiar?_____

11. What is(are) the image(s) of the new symbol(s) that have been
    introduced?_____

12. Does the mirror help you identify the base? _____
    How?_____

SELECTED LEARNING EXERCISES

1. Which of these Post Lesson Activities would you use first with your students? Justify your choice.

2. Which of these Post Lesson Activities do you think would be most difficult for students to complete? Why? What kinds of readiness would you provide to prepare students for this activity?

3. Develop a worksheet for one of the embodiments (either weights or money) that is designed to develop addition skills.

4. Describe three other physical embodiments that you might use to develop numeration systems. (Hint: use toothpicks, bundles, match boxes, kitchen match boxes, cigar boxes, etc.)

5. What students do you think would require another sequence of parallel lessons? Would you use this third set of lessons immediately or wait awhile? If you decide to delay these parallel lessons, how long would you wait?

6. Which of the four Post Lesson Activities would you use for assessing students' ability to transfer ideas about numeration systems?

7. Small group or class project. Select a mathematical topic and describe different physical materials that could be used to develop the concept. Identify specific ideas to be developed or objectives to be reached and then construct two series of parallel lessons that reflect the objectives.

8. Small group or class project. Use the first two lessons on numeration systems as a guide to develop fundamental ideas about trading with second, third, and fourth graders. Orally explain the trading rules and then provide students with some related exercises. When this demonstration lesson has been completed, summarize the results and your reactions. In particular, did the second graders respond in a manner similar to the fourth graders? Did the embodiments have about the same difficulty level? Was the explanation of the trading rules clear to the students? Were the related exercises appropriate for these students? How would you follow up this activity?

BIBLIOGRAPHY—CHAPTER 7

Bradley, R. C. and Wesley N. Earp, "The Effective Teaching of Roman Numerals in Modern Mathematics Classes." *School Science and Mathematics*, 66: (1966): 415–420.

Chandler, Arnold M., Editor. *Experiences in Mathematical Ideas*, Vol. 1. Washington, D.C.: National Council of Teachers of Mathematics, 1970.

Davidson, Patricia et. al. *Chip Trading Activities-Set 1*. Newton Lower Falls, Massachusetts: Growth Activity Products, 1971.

Dienes, Zoltan P. *The Arithmetic and Algebra of Natural Number*. Harlow, Essex: Educational Supply Association, 1965.

Lerch, Harold. *Numbers in the Land of Hand*. Carbondale, Illinois: Southern Illinois Press, 1966.

8 TEACHER MADE ASSIGNMENT CARDS AND LABORATORY LESSONS

IMPLICATIONS FROM A COGNITIVE MODEL

Teachers interested in more student involvement and unable to purchase commercially available laboratory lessons (see Appendix C), frequently ask about the possibility of developing their own. This chapter provides some guidance toward this end. The suggested technique is not difficult to master. Furthermore, you will find your expertise increasing rapidly as you continue to gain experience. Before turning to the specifics of this technique let us look briefly at the rationale for its existence.

In his book, *Toward a Theory of Instruction*, Jerome Bruner* identified three temporally ordered stages or levels of representational thinking: the enactive, the iconic, and the symbolic.

The Enactive Level as its name implies, is in evidence when the individual is physically active; manipulating, constructing or arranging elements of the real world. The crucial identifying factor for the enactive level is *first hand* interaction with the physical world. This level is the most concrete and is the stage at which the rudiments of concepts are formed. Concepts at this stage are present only insofar as they relate directly to the real world. They will, if properly nurtured at some later time, be abstracted and utilized by the individual in the symbolic mode.

Iconic means "picture, image or other representation." The iconic level as defined by Bruner is the second stage of representational thinking and is easily identified by the fact that it represents in descriptive verbal, pictorial, or some other form, happenings in the real world.

The Symbolic Level is the most sophisticated and is appropriate only after the individual has had experience with enactive and iconic representations of a given concept. This level is character-

---

* It should be noted that Jerome Bruner is located squarely in the cognitive camp from a philosophy of learning point of view. Although minor differences persist, he is in substantial agreement with Jean Piaget and Zoltan Dienes as to the conditions most likely to optimize learning.

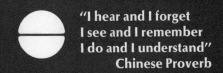

ized by the manipulation of symbols irrespective of their enactive or iconic counterparts.

An example may clarify these distinctions. Consider the topic: place value and the addition of 2 digit numerals; the specific problem being 17 + 34.

The child approaching this concept at the enactive level would be found setting the problem up on an abacus, using a place value chart, a pile of toothpicks or straws, or some other physical embodiment of the problem situation. The actual re-grouping would be done by the child. If he were using straws, he would join 7 straws with 4 straws, realize that these could be regrouped into 1 bundle of 10 with 1 straw left over. This in turn could be combined with the 4 bundles of ten (1 + 3), yielding a final result of 5 bundles of ten with 1 straw left over.

The child approaching the same concept (17 + 34) at the iconic level would be found looking at a picture(s) of the operation just described. He would probably be asked to supply the answer after observing a series of drawings describing or detailing another's actions. i.e.,

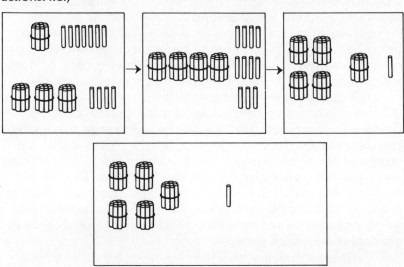

Finally the symbolic level would be:

$$\begin{array}{r} 17 \longrightarrow 10 + 7 \\ + 34 \longrightarrow 30 + 4 \\ \hline 40 + 11 = 51 \end{array}$$

or, in the final analysis,

$$\begin{array}{r} 17 \\ + 34 \\ \hline 51 \end{array}$$

Notice the child is interacting with the real world only at the enactive level. The iconic level is appropriate after the child has had the first hand experience afforded by his "enacting" the actual problem situation. Success at the symbolic level is likewise dependent upon satisfactory experiences at the previous two stages. Note the complete lack of physical or real world referents at this level. What is required here is that the person involved realize that these symbols are merely a shorthand method of representing an actual occurence in the real world. Surely this cannot happen unless the individual has actually encountered these experiences.

Textbooks, by their very nature, are "forced" to approach the teaching of mathematics at the iconic or symbolic level. That is, the textbook continually represents—using photographs, line drawings, diagrams or some other means—real world happenings. It tacitly assumes that observation of these happenings to be a sufficient condition for effective mathematics learning. This is not undesirable *unless* use of the textbook precludes essential prerequisite enactive experiences. Unfortunately, for a variety of reasons, this is the case. Lack of time, lack of materials, external pressure to complete the textbook, fear of an intolerable noise level, and pressures to have students do well on end-of-year standardized tests are among the reasons often given for eliminating the enactive experiences in the mathematics program. The authors feel that such omissions in the early part of the child's mathematics program are accountable in large measure for the massive remediation programs which are commonly found beyond the first or second grade. In a very real sense the system forces the slow-learning child to abstract before he is ready, that is, before he has an adequate concrete conceptual base from which to abstract, and as a consequence, he is unable to keep up with the rest of the class. Rather than concerning ourselves exclusively with large scale efforts to remediate those children who have failed the system (or more appropriately, those children the system has failed) it would be well to begin to consider the reasons for that failure. To begin to be concerned with the prevention of failure seems to us to make at least as much sense as being concerned with remedial techniques to be employed after the fact. It is interesting to note that remedial techniques are usually enactive in nature, requiring the child to interact in some way with the real world. Perhaps if he had done that to begin with he would not be in need of remediation at some later time.

The mathematics laboratory can now be thought of as a

vehicle to reassert the essential nature of the enactive experience in the mathematics program. How does this relate to the theme of this chapter? (You may recall the theme as being "Teacher Made Assignment Cards and Laboratory Lessons.") The message is simply this: you as the classroom teacher can develop laboratory lessons for your own classroom directly from the textbook you are currently using. *The task in reality reduces to little more than transforming a problem given in the iconic or symbolic mode into the enactive format.* That is, instead of having students read about another's actions and analyzing someone else's data, rewrite or reconstruct the situation so that your students are actually involved in the physical setting depicted by the textbook problem. This will find students experimenting and analyzing results which are now more relevant to them since they are actually involved in the process of generating those data.

Not all textbook "problems" lend themselves to such transformations. Professional judgement should certainly influence your decision to "enactivate" a textbook situation. For example, if a textbook problem dealt with the amount of soap consumed by a window washer in midtown Manhatten in a day's time (given the ratio of soap consumed per unit of window surface area and the size and number of windows washed in an 8 hour period), it would probably be unwise to locate your class on scaffolds attached to the local skyscraper for the day's mathematics activity.

There are however, many situations found in any commercial text which lend themselves very nicely to such a transformation. Consider for example, the wide variety of textbook problems dealing with the construction of bar graphs, which incidently, can be found at almost any grade level. Rather than asking children to make a graph of the following situation

NAME	Fred	Mary	Joe	Bob	. . . .	Ruth
WEIGHT	88	73	100	97	. . . .	78

why not borrow a scale from the nurse's office or bring one from home and construct a histogram (bar graph) of the weights of your class? The latter activity is guaranteed to generate more interest and enthusiasm because each child now has a real contribution to make to the class activity. It is real (as opposed to contrived) to him and he is likely to see more purpose in participating. Note that the children are learning basically the same mathematics; it is the initial delivery system which differs.

It would be fruitless to attempt to detail every conceivable mathematical situation which could be transformed into a laboratory experience. The transformation is not a difficult process. You will find yourself becoming more adept with each attempt. Try several of the exercises below to convince yourself of this. Before you do however, let us offer some final cautions and remarks.

1. Be sure that directions and questions to students are concise and

clear. If students are to proceed on their own they *must* know precisely what is expected of them.

2. All materials needed should be presented to the student with the laboratory lesson *or* provide a succinct description of where these materials can be procured.

3. If equipment needs to be assembled (e.g., balance beam), be sure to include appropriate instructions.

4. If pupils are ability grouped, occasionally, for laboratory lessons the top group may only require one enactive experience after which they may proceed to further practice with problems in the textbook (iconic). The more concrete oriented children may spend several days doing various enactive mode activities and only slowly proceed to the iconic representations of the same type of problems under the teacher's careful direction.

SELECTED LEARNING EXERCISES

Listed below are exercises similar to those found in commercial mathematics texts. Choose a partner and discuss modifications which would need to be made (materials needed, class organization procedures, methods of analysis, etc.) to transform these exercises into laboratory experiences. Make a list of extensions which you feel to be appropriate.

1. The table lists the number of absences in one week at Main Elementary School.

Day	M	T	W	Th	F
No. of Absences	23	41	19	20	24

Make a histogram of this information.

2. Jack threw a pair of dice 100 times and got the following results:

Sum of Two Dice	1	2	3	4	5	6	7	8	9	10	11	12
Frequency	0	2	5	8	11	14	19	15	10	9	4	3

Make a histogram of his results. Why did he not get any 1's?
What percent of the time did he throw a 7? 6 or 8? 2 or 12?
If he repeated his experiment would you expect the same results? Why or why not?

3. Mr. Smith and his family are planning their summer vacation. He plans to spend 20% of his money for transportation, 30% for lodging, 25% for food, 15% for incidentals and the rest for entertainment. If he begins his vacation with $300, how much does he spend in each category?

4. The volume of a rectangular solid divided into smaller units is found by multiplying the number of units along the base, by the number of units high, by the number of units deep. i.e.,

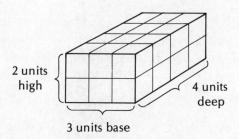

2 units high

4 units deep

3 units base

The volume of this rectangular solid is $2 \times 3 \times 4 = 24$ square units. Can you find the volume of the following rectangular solids?

4a.

4b.

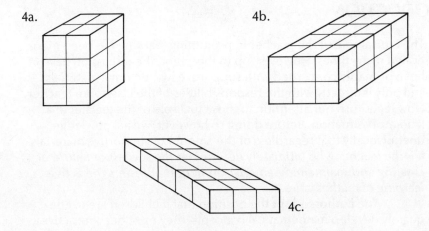

4c.

5. If a 57 foot telephone pole falls across a cement highway so that $16\frac{3}{6}$ feet extend from one side and $14\frac{9}{7}$ from the other, how wide is the highway?

This last example is unfortunately all too indicative of the type of trivia with which much of the elementary school mathematics program is concerned. Although it can be easily transformed into a real world experience, the authors are convinced there are better ways to spend the students' time and include it as an example here primarily to emphasize this point.

BIBLIOGRAPHY—CHAPTER 8

Bruner, Jerome. *Toward a Theory of Instruction*. New York: W. W. Norton and Co., 1966.

Charbonneau, Mannon P. *Learning to Think in a Math Lab*. Boston: National Association of Independent Schools, 1971.

Lorton, Mary Baratta. *Workjobs*. Menlo Park, California: Addison-Wesley Publishing Co., 1972.

9   ROLE OF THE TEACHER

INTRODUCTION

The crucial role of the teacher in promoting learning has been mentioned many times. However, up to this point, the discussion has been primarily concerned with supportive instructional materials and only indirectly with the responsibilities of the classroom teacher. Now let us turn our attention to the actual role of the teacher in a laboratory situation. Before doing so however, we acknowledge unequivocally that regardless of the amount of supportive materials, *it is the teacher who ultimately determines the success or failure of creating and maintaining an environment that fosters the active learning of mathematics.*

The authors adopt the position that if children are to adequately develop mathematical concepts, they must become actively involved in experimental situations within which these concepts are embodied. These experiences provide the sensory data upon which mathematics learning and, ultimately, mathematical abstractions are based. The role of the teacher in such a setting should be characterized by sound pedagogical techniques. In particular this requires the provision of a learning environment in which students can, within certain general parameters, experiment and try things for themselves. It is, of course, up to the teacher to identify these parameters.

Piaget stated his perceptions of the relationship between experiences and learning when he said:

> If they read about it, it will be deformed as is all learning that is not the result of the subject's own activity . . . teaching means creating situations where structures which may be assimilated at nothing other than a verbal level . . . a teacher would do better not to correct a child's schemas, but to provide situations so he will correct them himself. . . .[1]

A similar view was expressed by Dienes when he commented

---

[1]   Eleanor Duckworth, "Piaget Rediscovered," *The Arithmetic Teacher* 11:498 November, 1969.

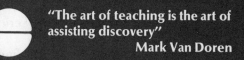

"The art of teaching is the art of assisting discovery"
Mark Van Doren

upon the relationship between the role of the teacher and an active approach to the learning of mathematics.

> It will probably be necessary to abolish almost completely the present method of class teaching with the teacher pontificating from a central position of power and to replace this by individual learning in small groups, from concrete material and written instructions, with the teacher acting as a guide and counselor.[2]

Thus the teacher's role becomes that of facilitator of learning: the one who provides experiences so that others may learn.

Obviously planning an active learning approach to mathematics is much different than relying solely on a textbook oriented mathematics program already in existence. Such planning requires 1) a thorough knowledge of individual pupils, including background, skills and ability; 2) a broad selection of supportive mathematics laboratory resources; 3) an extensive supply of activity oriented lessons; 4) careful organization of classroom, pupils and activities; and 5) efficient and effective use of teacher time. In short, implementation of this scheme of learning requires the best efforts of a true professional.

SUGGESTIONS FOR A BEGINNING

How can a teacher initially establish a laboratory approach in the classroom? Although this question is frequently asked, the answer is neither simple nor unique. In fact, the number of different answers is at least equal to the number of teachers who have developed an activity oriented learning mode in their classrooms. That is to say, there is no one best way to get started. The procedure is usually determined by each teacher's unique situation. Nevertheless, there

[2]   Zoltan P. Dienes, *Building Up Mathematics* (revised edition; London: Hutchinson Educational Ltd., 1969) p. 17.

are several ideas worthy of serious attention by teachers considering a change.

The following guidelines are not proposed as being exhaustive, nor is any hierarchy of importance suggested by the order in which they are discussed. It is felt, however, that each of these are necessary before such a learning environment can be established.

The teacher must, first and foremost, genuinely support an active approach to learning. At the onset, skeptics, including other teachers, parents, and some administrators, will probably outnumber the proponents of this mode of learning. As a result the major burden of proof lies on the innovator.* If this implementation period becomes a fiasco, skeptics will probably not hesitate to say, "I told you it wouldn't work." On the other hand, the gradual development of an effective activity oriented program will frequently convert earlier skeptics into bona fide supporters.

As with any innovative approach to learning, one can expect problems at the initial stages. These problems may vary widely in both importance and frequency of occurrence. For example, early problems may range from "securing enough paper cups for an activity" to "convincing the school board of the soundness of this mode of learning." In addition to overcoming initial difficulties the teacher must be willing to assume a more demanding role in the classroom. For example, initially it will take more teacher time to develop and organize an activity oriented program. It will require that the teacher often assume the role of the "devil's advocate" by asking questions rather than the more conventional role of providing answers. The ability to ask questions presupposes a certain breadth and depth of knowledge in the subject discipline. Given the foregoing constraints, it is suggested that only those teachers who genuinely support this active approach to mathematics learning should attempt large scale modifications within their classrooms.

Support from the school administration, including the principal and other appropriate personnel, is essential. The rationale for this mode of learning should be clearly presented to everyone concerned. Likewise there should be a frank discussion of potential problems that would require administrative understanding and support. For example, in the activity oriented program students will be doing different things at different times. Some students will be completing one activity, while others are starting another. Some students will be moving about the room and some will be discussing an assignment. The arrangement of classroom furniture will vary according to the activity, and the noise level in the room may be high. This is considerably different from the image projected by the traditional classroom, and may cause some degree of alarm. Assuming such an

---

* It is imperative that some type of "support base" be established within the school concurrent with the implementation of the laboratory approach to learning. This base will hopefully include representatives from the administration and other members of the school faculty.

occurrence, the principal must be aware of the individual's efforts and be willing to support them to the best of his ability. This will necessitate careful pre-planning and the establishment of viable communication channels between all those who are likely to become involved.

The implementation of activity-oriented lessons should be on a small rather than a grandiose scale. This suggestion is based on several factors. It takes time for students to adjust to a different learning atmosphere. Students who have grown accustomed to assuming a "passive" role as learner, will need to adapt to a new learning mode whose success is dependent upon his active participation. It takes time for students to learn to operate effectively in group learning situations. Initially there may be a "felt need" by some students for the teacher to provide direct and complete group leadership. However, once the teacher has made it clear that group leadership must come from within, students will tend to wean themselves from teacher domination. It takes time for students who have become highly dependent upon the teacher to operate independently in a meaningful learning situation. It takes time for students to learn where materials and resources are stored. It takes time for students to effectively operate in a learning environment that requires them to collect, organize, and analyze empirical data. It also takes time for the teacher to procure, develop, and organize a reservoir of activities and the necessary supportive resources.

One proven procedure for beginning is the gradual introduction of activity oriented lessons with selected children. As these students learn the ropes, the teacher might then extend these activities to other children in the class. This gradual development is much preferred over an all at once approach. The former provides time for continuous adjustment and modification, whereas the latter (besides being unrealistic) invites chaos.

CLASS ORGANIZATION

In the activity oriented classroom, children progress from one activity to another. The range of learning experiences, as well as supportive materials, will be far greater than in the traditional classroom and will generally demand student involvement of varying degrees. Sometimes the activities are designed on an individual basis while other times some form of grouping is required. The size and composition of a group depends on several factors, including both student age and the nature of the activity.

It is, of course, unwise for us to make formal recommendations regarding specific group organization. There are no generalizations that can be safely made regarding the relative efficiency of any particular organizational scheme. Nevertheless, there are some ideas that warrant the serious consideration of the laboratory teacher.

Determining the proper group size for a laboratory activity is a problem regularly faced by the classroom teacher. There are, of course, many options open to the teacher in formulating groups. One of the most commonly used grouping schemes is to allow students to work in pairs or small groups. This provides much first hand experience as well as an opportunity for discussion of problems. In some activities, one or two children may perform an experiment while another records the data. After organizing the data collected, the students discuss and interact as they analyze their findings and propose solutions. Primary pupils seem to operate most effectively in relatively small groups. Although some intermediate grade children can function well in larger groups, perhaps four or five, others may prefer to learn on their own. As children become more mature they tend to become increasingly independent. This natural desire to assume independence is both fostered and cultivated in a laboratory setting. The teacher is constantly confronted with forming new groups of pupils for different activities. It is only through experience that one can find the optimal group size for a particular activity and for a particular group of children.

It should be recognized that group membership often determines the success or failure of a learning activity. There are a number of different grouping methods that might be used in a laboratory setting including friendship, mixed ability, or common ability schemes. Friendship grouping can result in an excellent learning experience, as the group attacks the activity with enthusiasm and candidly discusses the results. On the other hand, grouping by friendship can be unproductive and characterized by idle chatter when the group's attention is diverted or interest wanes.

Heterogeneous grouping can also be used effectively in the laboratory setting. The brighter children often assume a leadership role answering activity related questions raised within the group. Such children often learn more from the activity when they are asked to explain the "why" to others. The child of lesser ability will also benefit from an explanation provided by a peer, rather than from the teacher. Perhaps the greatest fear of the mixed ability grouping is that brighter children will dominate the activity, thereby intimidating other group members.

Homogeneous grouping can also be very effective for some activities. The big problem lies in the formation of groups. Students' learning patterns are often too complex to permit precise diagnosis. Thus, it is difficult to accurately assess pupil ability. The problem is further compounded by the fact that pupil characteristics are constantly changing. Therefore, if homogeneous ability grouping is used, the groups should not be permanent. Rather, they should be subject to continual review and appropriate modification.

The discussion suggests that each grouping scheme has its own advantages and disadvantages. Above all else, one should not become dependent upon a single method. It is often advisable to change the group structure as new activities are undertaken. The frequent formation of new and different groups can result in student growth in both the academic and social domains.

SOME SUGGESTED TEACHER DO'S AND DON'T'S*

The new role of the laboratory teacher has been frequently cited. Several important aspects have already been identified. It would, of course, be impossible to construct a comprehensive list describing characteristics of the ideal teacher in an activity oriented program. Nevertheless it is felt that some specific "do's and don't's" for teachers in an activity oriented program are both appropriate and essential. Recognizing the possibility of omitting important facets of the new role, the authors nevertheless suggest the following items for consideration.

1. *Do construct activities that provide for Multiple-embodiment of the concept.* It is difficult, and often foolhardy, to abstract or generalize from a single experience. Thus the pupil should be presented with different situations manifesting the concept or structure to be learned. For example, in developing the concept of three, children might examine sets with three elements for one activity. The number line, balance beam and Minnebars might also be used to provide different "embodiments" for the same concept. The case for "multiple embodiment" has been presented earlier. Although the idea is pedagogically sound, it has yet to receive widespread use by classroom teachers.

2. *Do provide a wide range of activities.* This is a sequel to providing "Multiple-embodiment" of a concept which requires the existence of parallel lessons or activities that focus on one mathematical concept. It also demands the existence of sufficient series activities to insure a broad scope and continuous sequence of mathematical ideas. The teacher has the primary responsibility for the procurement and development of this wide range of activities which will contribute substantially to the success or failure of this approach to learning.

3. *Do prepare in advance for the activity.* Be sure you, as the teacher, use the materials before they are actually used by the pupils. As you make this "trial run" you should consider questions such as: What prerequisite skills are needed before these materials are introduced? Are the directions clear and can they be easily followed? Are there an adequate number of leading questions? Are the activities commensurate with the level of

---

* It is with deep reservation that the phrase "teacher do's and don't's" is used. The authoritarian tone of this phrase smacks of heresy; however, the intent of this section is to briefly summarize some of the more salient factors that will determine the success or failure of an activity oriented program. Many of these suggestions are discussed elsewhere at greater length.

pupil ability, and do they provide an appropriate embodiment of the mathematical concept? What are potential problem areas, and how might they be avoided?

4. *Do prepare the students.* The type of preparation depends on both the materials being used and the age of the students. Above all else, be sure students are ready to profit from experience with the materials. Care should be taken to provide the necessary directions for beginning the activity. One must guard against telling students precisely what to do. On the other hand, sufficient direction should be provided to prevent mass confusion, which may quickly lead to discipline problems.

5. *Do provide for individual differences.* Careful consideration must be given to the selection of activities to insure that they are commensurate with students level of development. This requires that a wide range of both "Parallel" and "Series" lessons be available, with the teacher prescribing the type and nature of the activity most appropriate for the student at a particular stage of development.

6. *Do prepare the classroom.* Check to insure that all required materials are on hand. Also be sure they are operative, accessible and available in sufficient quantity. The arrangement of the classroom furniture should be considered to assure that it will facilitate the planned activities.

7. *Do encourage students to think for themselves.* The use of manipulative materials in an informal situation provides an ideal climate for creativity, imagination and individual exploration. This atmosphere encourages students to think for themselves. However, in order to get students to begin, and then continue to think for themselves, it is imperative that the teacher provide encouragement and show respect for student's ideas. A teacher's dismissal of a student's contributions as being trivial, incorrect, worthless, stupid, crazy, etc., will repress future ideas.

8. *Do encourage group interaction.* Discussion within, as well as among, groups can be intellectually stimulating. Encourage students to communicate with their peers and teacher. Oral expression of ideas is an excellent means of improving communication skills. The importance of having this opportunity to express one's thoughts, observations and ideas cannot be overestimated. As students grow older this freedom to express personal ideas is accompanied by a responsibility to defend or at least support a position, should the need arise. Some teachers fear that one student will dominate a group of peers. This may sometimes happen; however, the careful selection of group membership can keep this problem to a minimum.

9. *Do ask the students questions.* It is often essential that certain points be called to the students' attention. Sometimes big ideas are missed completely. Other times one child may divert group attention to some minor or obscure point. It either case, you must be prepared to ask pertinent leading questions.

10. *Do allow children to make errors.* Some may view this as heresy. However, children must have an opportunity to be wrong or make a mistake. Often greater learning and more lively discussion follow an incorrect answer than a correct one. Besides, the natural learning process is characterized by much trial and error learning. To do otherwise, that is to attempt to eliminate incorrect answers or faulty speculation, is to create a highly artificial learning situation.

11. *Do record observations.* In moving from group to group, make mental notes of pupil progress. Ask questions and listen to the answers, being careful not to assess the progress of the entire group by one student's response. Always be alert for clues that suggest the need for new and different activities. In order to preserve the freshness of student's comments and your own observations, you must take time to record comments while they are fresh in your mind. Otherwise many valuable bits of information will be lost. Immediate recording of observations places an awesome burden on the teacher. Nevertheless, it represents a vital function of the laboratory teacher, who must ultimately recommend follow-up activities. Wise recommendations can only result from continuous scrutiny of students actively involved in a laboratory setting.

12. *Do provide follow-up activities.* Discussion, correlated readings, reports, and projects, as well as replications of activities, enhance the prospects of learning. Searching questions forcing students to further analyze and synthesize their results can be very productive, as they encourage students to "pull together the loose ends." This might be followed by additional questions which require extrapolation from these activities and encourage speculation on the outcome of other related events.

13. *Do evaluate the effectiveness of materials and activities.* After each lesson, it is wise to make an assessment of its effectiveness. Immediately upon the completion of an activity it can be very helpful to note particular problem areas, strengths, weaknesses, suggestions, and define areas of needed improvement as well as possible areas of modification. A continuous re-evaluation of materials and activities ulitmately results in better use of better quality equipment.

14. *Do exchange ideas with colleagues.* Many new functions of materials and activities result from actual classroom use. Sometimes students either consciously or unconsciously propose additional uses. At times informal exploration with materials by either teacher or children suggests new activities. In any case the classroom teacher has added to the reservoir of potential uses for this set of manipulative materials. A mutual exchange of ideas among teachers allows each to profit from the other's experience. One of the authors is reminded here of the old addage "If I have a dollar and you have a dollar and we exchange dollars, then we both still have one dollar. But if I have an idea and you have an idea and we exchange ideas, then

you have two ideas and I have two ideas." We heartily recommend the sharing of ideas with your colleagues.

Now for some teacher don'ts!

1. *Don't use instructional materials indiscriminately.* Care must be taken to assure that these materials properly embody the mathematical concept being developed. Be sure the materials and the intended concept are in keeping with your objectives and the student's level of development.

2. *Don't make excessive use of manipulative materials.* Manipulative materials should be used only when they represent an integral part of the instructional program that could not be achieved more effectively without the materials. One exception to this might be manipulative materials that are directed more toward recreation. There are instances where the traditional curriculum fails to reach many students. Often the recreational aspect of manipulative materials has attracted the attention of these youngsters and eventually paved the way to more academically oriented activities. Some teachers fear that excessive use of manipulative materials will lead to overdependence on physical representations. There are cases where the manipulative materials are used as "crutches." However, most students will phase themselves away from the materials when they are no longer in need of them and have reached a higher level of intellectual development. Excessive use of manipulative materials is generally characterized by children showing signs of boredom. This doesn't always mean the materials should be removed, however; it suggests the need for raising additional questions or extending the concepts being explored.

3. *Don't hurry the activity.* Once the concept has been developed, most children are eager to explore other ideas. However, every student should have ample opportunity to complete the activity, thereby convincing himself of the principle or formulating the concept. Hurrying through the activity may impose unnecessary pressure on some pupils as well as creating an artificial learning situation. Few individuals learn well when they are rushed. Individual differences, being what they are in most classes, means that some children may formulate the concept within minutes, whereas other children may require several days or perhaps even longer. Rushing children through activities does not solve the problem, but rather compounds it.

4. *Don't rush from the concrete to the abstract level.* This is a sequel to the previous suggestion. Perhaps the most frequent error in using activity oriented lessons is the speed at which children are rushed from the concrete stage to the symbolic level. There seems to be some myth that you are not doing mathematics unless you are actually working with symbols. This, of course, is not true. Most mathematics at the primary level should be done without symbolization. In fact, if serious

consideration were given to Piaget's research, nearly all mathematics in the primary grades would be designed to accommodate the Concrete Stage of Intellectual Development. In an effort to pinpoint the problem, it must be remembered that the ability to think abstractly (one aspect of which is symbolization) occurs quite late in concept formation. Symbols are reserved for describing or making a record of the concept or mathematical principle. Hence, they can only be properly used after the concept has been abstracted. Since the process of learning a mathematical abstraction is time consuming, it is ludicrous (at least with most elementary children) to use manipulative materials one or two days, and then move directly to the symbolic level. Such use of manipulative materials can only lead to confusion, since few children will abstract the concept or mathematical principle. The wrong kind of experience may result in children viewing manipulative materials as toys or entertainment, and in no way related to mathematics.

5. *Don't provide all the answers.* In working with activity oriented materials pupils acquire experience in abstracting from a set of phenomena, or a body of data. As each child is actively involved in this process, conflicts frequently arise. One pupil has one answer, another child has a somewhat different result. Often the first reaction of the teacher is to settle the issue by providing the correct answer. It is difficult to resist the temptation to tell the correct answer, but resist the teacher must do! To do otherwise is to discourage individual pupil thought, squash natural curiosity to search for other solutions, promote pupil dependence on teacher rather than pupil independence, and preclude further discussion of the problem, as everyone now knows the correct result. On the other hand, you may decide to ask some leading questions; you may have the pupils explain their solutions; you may wish to have them replicate the activity using the manipulative materials or pursue some other alternative. Regardless of the option selected, the teacher must refrain from serving as the "purveyor of truth," and source of all knowledge. Remember, that to children and adults alike, "the art of being a bore consists in telling everything."

The suggestions made represent a clear departure from the traditional role of the classroom teacher. The focus is changed from a teaching situation to a learning situation. Thus the responsibility for learning is borne by the pupil rather than the teacher. The teacher is not there just to teach, but to encourage learning. Although this new role is very demanding in terms of both time and energy, it is also very rewarding. After one has carefully provided a series of pupil oriented activities leading toward the development of a mathematical idea, nothing can give a teacher greater satisfaction than to see the look of triumph and excitement in a child's eyes when he has grasped an idea or discovered a relationship.

It should be clear that the responsibility assumed by the

student for learning does not in any way suggest that the teacher abstains from teaching. Unless the teacher provides pupils with materials, equipment and ideas, there is no reason to believe that significant intellectual development will emerge from the mathematics laboratory. Truly a new role is necessary for the teacher. In assuming such a role it should also be clearly understood that the activity oriented mathematics program places more, not less, responsibility on the teacher. In traditional oriented programs, most curricular decisions are dictated by the textbook. To a far greater extent, in the activity oriented program it is the teacher who must know how and what children need to learn, and when they need to learn it.

SELECTED LEARNING ACTIVITIES

1. It has been said that process oriented activities which have an experimental component can be used as a vehicle to facilitate mixed ability (heterogeneous) grouping patterns. The rationale proposed for this argument often contains reference to the fact that students are able to involve themselves in learning activities at varying levels of sophistication, and that experiments as such lend themselves nicely to the partitioning of student responsibilities or roles even within a single experiment.

    a.   Do you agree with this position?

    b.   Discuss the pro's and con's of this argument.

    c.   Develop or identify a laboratory lesson and indicate how students within a single group having varying abilities might partition responsibilities so that each is participating at a different level of sophistication but one that is appropriate to his ability level.

    d.   Share your answer(s) to c) above with another group.

2. Small group or class project.* Select a group of children and a mathematical topic that can be effectively developed within several days. This activity may be done with an entire class or with smaller subgroups of children. Construct two or three appropriate learning activities. Acquire the necessary materials, identify related exercises, develop leading questions, and do whatever else you feel to be necessary as you prepare to help children develop the mathematical concept. Select one or two members of your group to present the lesson to students. Other group members will serve as observers. Upon completion of the lesson, discuss significant happenings within the entire class as well as those observations made by members of your group. Also consider the following questions:

---

*   We feel that to propose this activity, in a setting other than an actual classroom would be fruitless or at best an academic exercise. Therefore, we urge you to try these exercises with children. One learns to swim by swimming. How would you suppose one learns to teach?

Did the activity embody the mathematical concept. Could and
should it be improved?

Is another activity or embodiment needed? If so, what activity
would you recommend?

Was advance planning evident? How?

How effective were the leading questions? Were they really
needed?

Was there interaction among students? between students and
teacher?

What follow-up activity would you recommend?

If you were to teach this lesson again, what would you do the
same? Do differently?

3. At this time do Exercise Number 8 from Chapter 7, if you have
not already done so.

BIBLIOGRAPHY—CHAPTER 9

Biggs, Edith and James MacLean. *Freedom to Learn.* Menlo Park, California:
Addison-Wesley, 1969.

Dienes, Zoltan P. *Building Up Mathematics.* (Revised Edition) London:
Hutchinson Educational, 1969.

Duckworth, Eleanor. "Piaget Rediscovered." *The Arithmetic Teacher*, 11:
(1964):496–499.

Reys, Robert E. "Considerations for Teachers Using Manipulative Materials."
The Arithmetic Teacher, 18:(1971):551–558.

Rogers, Carl R. *Freedom to Learn.* Columbus, Ohio: Charles E. Merrill Publishing Co., 1969.

10 EVALUATION

INTRODUCTION

Historically, the scope of evaluation in mathematics has been quite narrow as it has been focused almost exclusively on student mathematics achievement. Furthermore, indicators of mathematical progress have characteristically been dominated by reports of student attainment of basic computational skills. It should be recognized that many instructional processes are either overtly or covertly influenced by evaluation. Evaluation, particularly in a laboratory setting, must be concerned with the teaching process as well as with the learning process.

A dissertation expounding the merits—or lack thereof—of various evaluative techniques transcends the nature of this discussion. However, since evaluation plays a fundamental role (and rightly so) in today's schools, it does deserve special attention. It is recognized that every good teacher is constantly evaluating individual student progress as well as the effectiveness of the daily instructional activities. It is also recognized that evaluation is a complex time-consuming task that requires both technical proficiency and professional judgment. Furthermore, it is acknowledged that there is a need for evaluation in the mathematics laboratory just as there is in the regular classroom.

Evaluation of student progress is broadly conceived by the authors as being based on both objective and subjective data. There are many evaluation activities that fall within the domain of objective evaluation. Traditionally, the single activity having the greatest influence on the evaluative process in the public schools is the formal test. As indicated earlier, most mathematics tests place a heavy premium on computational facility either overtly (drill type exercises) or covertly ("problems" emphasizing or requiring speed and accuracy in computational endeavors). Although the appropriateness of such tests is suspect in the regular mathematics class, it must be seriously questioned in the laboratory setting. Good mathematics teachers use a variety of evaluative techniques, some of which include: achievement, oral,—one to one and group discussion—take home, diagnostic, open book, and performance tests. Although each of these tests may be effectively used by the laboratory teacher to evaluate student progress, it is the performance test that is perhaps uniquely suited to the laboratory approach and therefore warrants special consideration.

PERFORMANCE TESTS

Since in the laboratory setting students are actively involved in the learning of mathematics, it seems reasonable to use similar means to measure mathematics achievement—performance tests that are based on physical involvement in a problem situation. That is, a student may be given a set of physical materials and asked to respond to specific questions, the answers to which can be derived from direct interaction with the materials. In some cases, the questions may not *require* actual use of the materials, but they are available if needed. Thus the pupil decides what materials he wants to use, if indeed he wants to use any, as well as when and for how long he wants to use them.

A performance test may require a pupil to discover a new relationship through manipulation, experimentation, measurement, drawing, or, investigations of patterns. Thus the performance test provides a means of assessing ability to discover patterns, identify relationships, collect and organize bodies of data and ultimately to solve "real" problems. With a performance test each question, or set of related questions, is normally located at a separate "testing station."

Other embodiments of the same mathematical concept can often provide the bases for a performance test. Any of the four Post Numeration System Activities in Chapter 7, for example, could serve as a performance test over numeration systems. One could also extract portions of these Post Activities which are related to specific properties, and use them as a performance test.

In some cases, the performance test is embedded in the laboratory lesson, providing for evaluation of progress while the pupil is still involved in the activity. Figure 1 shows one set of questions related to a performance test entitled "Area and Perimeter of Polygonal Regions." This illustrates how a performance test might be integrated into an actual lesson, as opposed to a culminating activity as in most traditional testing situations.

Figure 1.

Topic: Area and Perimeter of Polygonal Regions

For this lesson you will find some one square unit pieces of cardboard.† Take the suggested number of square units and arrange them so as to form a rectangular figure with maximum perimeter. (Remember you must use the required number of pieces, and you cannot stack the pieces on top of one another or allow them to overlap. No empty space may be left in the interior of the figure.) Then, rearrange them so that you have a figure with minimum perimeter. Take the suggested number of square units and complete this table:

Number of Square Units	Minimum Perimeter*	Maximum Perimeter*
6		
9		
12		
16		
20		
25		

*Sketch the figures with the minimum and maximum perimeters on this graph paper.

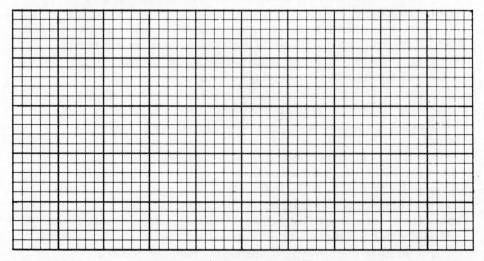

† Provide at least 25 square unit pieces at each station.

After you have completed the table and sketched the figures, answer these questions:

1. If you had 40 square units, what do you think would be the maximum perimeter?_____ the minimum perimeter?_____

2. If you had 100 square units, what do you think would be the maximum perimeter?_____ the minimum perimeter?_____

3. What relationship(s) do you notice between the number of square units and the minimum perimeter?

4. What relationship(s) do you notice between the number of square units and the maximum perimeter?

The laboratory teacher may also set up a number of separate testing stations in the room when evaluation is deemed necessary. Usually, however, all children are not ready for the same test at the same time. It is, therefore, more appropriate to set up performance tests focusing on certain concepts for individual pupils. Children are requested to take a performance test only after completion of the prerequisite activities. This means that some pupils may take a given performance test days, weeks, or even months before other pupils would be ready for the same test. Because of the individualized nature of the mathematics laboratory it is also probable that all students will not be evaluated using the same instruments, since all students will not have been involved with exactly the same mathematics activities.Such a testing procedure requires more teacher time and organization to administer. Although this practice is very demanding, it provides a testing situation that is more in keeping with sound pedagogical and evaluative procedures.

RECORD FORMS FOR STUDENT PROGRESS

In order to provide student progress information to both the learners and the teacher, it is recommended that a brief narrative record of each activity be made. Information can be recorded on cards (4" × 6") to facilitate handling and filing. Each card provides basic information such as:

1. name of student
2. name of activity
3. starting and ending dates
4. comments by teachers
5. recommended follow-up activity, see (Figure 2)
6. cassette recorder information (If appropriate)

Although some information can be recorded only when the activity has been completed, it is important that relevant anecdotal records be kept where possible and/or appropriate. Prompt recording of observations provides a breath of freshness and spontaneity to the report that is otherwise lost. Waiting until the lesson is completed to record all observations may delay students in a learning activity as some clarification of details is often required to get the child back on the right track. Delayed recording will also likely result in omitting significant items and in losing some of the individual flavor of the report. In such cases the personal report is often replaced by a stereotype commentary which is dominated by general comments and cliches that are not nearly as meaningful.

Initially this procedure of recording annotations will take more time and this fact should be accepted. As teachers gain experience in observing and recording information, ingenious methods to effectively perform this function will no doubt be developed and refined. One of the current advantages of this procedure for elementary teachers is that their comments and observations of children in a laboratory setting will cut across subject areas. Consequently problems of reading, listening, writing, etc., will be integrated into the evaluation of progress in mathematically related activities.

STUDENT PROGRESS RECORD CARD.

Student Name   Lesson or Activity

Date started   Date completed   Lesson Number

Areas of Strengths   Area of Needed Improvement

Annotated Comments: 1.

                            2.

                            3.

The next activity should be:

Lesson or Activity Name

Number

Cassette Recorder:
    Tape Number: _____

    Tape Footage: Before _____

                After _____

                         Teacher

Figure 2

There are, of course, many times when a teacher is unable to observe each child or group during a particular activity. In such cases, a brief discussion of the activity with the student can often compensate for firsthand observation. Such discussions should be characterized by a free exchange of ideas and questions between the student and teacher. Extreme care should be taken to assure that the discussion does not deteriorate to an interrogation, in which the student is faced with a constant battery of questions. Just as with firsthand observations, any notes based on this discussion should be recorded immediately.* Since many discussions with different children will occur daily, a teacher cannot afford to depend on memory to keep a permanent record of these observations.

Each card can provide information helpful to the teacher in making day-to-day educational decisions. As the teacher progresses, a cumulative file is established. This longitudinal record provides a capsule report of individual student progress in the laboratory. Among other things the teacher can readily determine which students are working on which activities and at which level. Furthermore, information concerning the number and kind of parallel lessons that each student has completed is also available. Such information provides an excellent perspective of individual student progress within the laboratory setting. It also provides for a direct assessment of areas of strengths and weaknesses which are invaluable in the prescription of new student assignments.

This cumulative file also provides an unparalleled opportunity for the teacher to identify changes in student behavior. A cursory review of a child's file may suggest areas of progress. For example, the child may have encountered no observable difficulty when studying modular systems. Yet the file may reveal that the same child has spent an unusual amount of time studying parallel lessons on numeration systems. Thus, observable changes in individual behavior, which might go undetected, are often gleaned from a review of the cumulative file. Of course, the file only brings evidence to the attention of the teacher, it does not provide a prescription to alleviate the problem. Once the review of the evidence has been made, it is the responsibility of the teacher to determine appropriate follow-up activities for the student.

The record card (Figure 2) and the cumulative   record file (if regularly used) provide much useful information to the teacher. There are of course other alternatives in gathering and organizing data to help assess pupil progress.

---

* A cassette recorder can be very helpful. Teachers can record personal observations as well as student-teacher consultations. These recordings can refresh one's memory concerning the way a student has completed a project and can be useful in detailing actual student responses and questions. When using this device, be sure to record the tape's footage both before and after the observation/interview for each child on his record form; this will permit efficient access to this information at a later time. (See Figure 2).

One procedure to streamline data gathering is a student-by-objective matrix similar to the one constructed in Figure 3.*

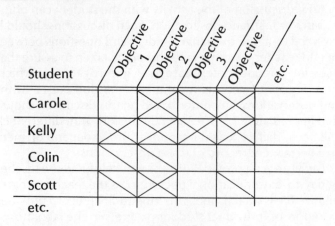

Figure 3

An empty box indicates no evidence for that student on that particular objective. A single slash ⬊ implies that some evidence is available, or that the student is working on that objective, but that the amount of evidence is incomplete, and it is not at this time possible to determine whether or not the student has accomplished the objective or mastered the skill, as the case may be. A double slash ⊠ implies that the student has mastered the skill or exhibits the desired behavior satisfactorily.

Thus in our example, Carole has mastered objectives 2 and 4, is working on objective 3 and no evidence is available regarding her progress on objective 1. Kelly has satisfactorily completed all four objectives. Colin is working on objectives 1 and 2 with no evidence available on objectives 3 and 4. Scott, on the other hand, is working on objectives 1 and 4, has completed 2, with no evidence available for objective 3.

A large amount of information can be quickly obtained from such a matrix. With a little practice one can easily gain proficiency with this procedure. Such information should be shared with the student on a regular basis, both as a method of highlighting perceived needs as well as providing positive feedback for his accomplishments.

Viewed vertically, the columns provide information on the various objectives. For example: which objectives have been completed by all or practically all students? Which have been completed by few or no students? It also reveals which objectives have been begun by many students but completed by few. Such a matrix does not explain why such conditions are occurring, but it may alert the teacher as to possible problem activities, questions and/or objectives.

---

* This model was developed by Professor Alan H. Humphreys of the University of Minnesota. The authors are grateful for his permission to include it here.

ROLE OF EVALUATION IN THE IMPROVEMENT OF INSTRUCTION

The individualized nature of activities in the mathematics laboratory requires that the teacher develop a more personal relationship with the child. The teacher needs to be able to observe and record (mentally or otherwise) the kinds of responses made by individual students. The teacher needs to recognize that within a certain group, one child may respond at an immature level, whereas another child consistently operates in a more sophisticated fashion. This process of evaluating individual strengths and weaknesses, as well as achievement and ability, should be constantly occurring in the laboratory.

As children are involved in an activity, the nature of their interaction with peers, as well as the teacher, will suggest individual levels of concept formation. The observant teacher is alert for clues which provide suggestions and ideas for follow-up laboratory activities. Thus students, who have yet to develop a given concept, may be directed to a parallel lesson, while those who have formulated the concept can progress to an activity in a different mathematical domain.

Constant student evaluation allows teachers to provide appropriate learning activities to individuals within various levels of concept formation. A concomitant result of close scrutiny of individual and group work in the laboratory can be the improvement of the software utilized. The record form cards can be used to secure information helpful in revising lessons and activities. After sorting cards according to activity, the teacher can examine each resulting pile. This review often results in specific suggestions for revision of a particular activity. Suppose, for example, that in reviewing a particular set of record cards the teacher observes similar questions or misconceptions raised by several different students. Experience suggests there may be something inherent within the lesson that distracts or misleads the student. Such information can be used to revise and improve the lesson in question.

Teachers need to constantly evaluate daily instructional activities. The alert teacher can obtain a wealth of information concerning the success or failure, strengths and shortcomings of lessons through verbal interaction with students. Such feedback information may come at any time and take various forms, including: 1) student comments, 2) questions the children raise, 3) the terminology used and abused, 4) the skills exhibited 5) the conjectures made, 6) the reasoning and thought patterns followed and 7) the interest and enthusiasm shown for the learning situation. The responsibility of analyzing and synthesizing these data rests with the teacher, who can then utilize this information to improve all aspects of instruction.

GRADING

Evaluation of student progress in a mathematics laboratory, as in any class, cannot be adequately reported by a single letter grade. Be*sides being a very "crude" report of student progress, a grade fails to identify the individual strengths and weaknesses which must be recognized (by both teacher and learners) if true learning is to result. Students in a mathematics laboratory cannot be graded in the traditional manner without harmful results. One of the keystones upon which active learning is based, is that students are provided with informal learning activities commensurate with their developmental level. Undue external pressures to complete a fixed number of lessons with a fixed error rate† is completely incompatible with the philosophy of active learning.

Within the laboratory setting, the student must have an opportunity to explore ideas, make conjectures, and be creative. Each of these will, at one time or another, result in students making errors and searching blind alleys. Since learning results from these kinds of "errors," we must give the child a chance to make mistakes. Placing evaluative marks on lessons or activities is one of the best ways of stifling individual thought, thereby creating a very artificial learning situation. Since grades purportly indicate degrees of success or failure, students will generally regress to "working for the grade." Too often this results in a pre-occupation with the quantity of activities completed, rather than attempting to do well, whatever activities are attempted. When grades become the primary motivational factor, the goals of a true learning situation are distorted. In addition, any such evaluation scheme makes a mockery of individual differences, as each child is pushed (or dragged) from one activity to the next, in an effort to keep him on some arbitrary schedule.

The record form allows the teacher to record on-the-spot observations throughout the year. These cards may be utilized to provide the teacher with a cumulative record of student progress as well as the teacher's own written observations. Such information

---

* Evaluation by traditional means of reporting letter grades is felt to be totally inadequate. There can be no compromise on this issue. To do otherwise is to sacrifice the freedom needed by the teacher to make necessary educational decisions. Teachers using this approach to learning must be free to assess individual student progress without being confined by an evaluative "straight jacket."

For teachers in schools using letter grades to report pupil progress, it is our hope that exceptions can be made, rules bent and red tape cut to allow teachers to report student progress in a meaningful manner. Any evaluation scheme short of providing individual diagnosis of student strength and weaknesses will adversely effect this approach to learning.

† In fact a new definition of error is required if one utilizes an active approach to learning. See Chapter 3 page 18 for a reflection of our thoughts on this matter.

could be complemented by additional data from other sources, such as ability, homework, etc., to form the basis for reporting progress to both students and parents. This less formal method of reporting student progress is much more in the spirit of active learning than traditional evaluation techniques.

Therefore the authors recommend that formal letter grades be replaced by teacher annotations of the student's performance in all* activities attempted. These comments would summarize student strengths and weaknesses, as well as being diagnostic (identifying specific weakness) and prescriptive (recommending specific measure to alleviate the problems) in nature. Such a written report would be much more valuable to student, parents, and teachers than mere letter grades.

PARENT-TEACHER CONFERENCES †

The actual reporting of student progress takes many forms. Such reports should be made to both students and their parents. Reporting student progress should not be limited to periodic communication between teacher-child or teacher-parents, but should provide for a continuous free flow of two-way (and even 3-way) communication. Many situations develop that require immediate attention and cannot be delayed for weeks until a formal conference can be scheduled. It is imperative that all parties (teachers, students, and parents) feel free to exercise this opportunity for free and open communication.

The teacher has the responsibility of reporting student progress as observed in the classroom. In so doing the teacher must be professionally honest in assessing strengths and weaknesses. It is essential that areas of needed improvement be identified, as well as those areas in which much growth has occurred. Some teachers tend to provide all parents with a glowing report of student progress, because they find it difficult to make a frank assessment of weaknesses. Perhaps they are also hesitant to "rock the boat." This does not solve any problems, but often creates false impressions for children and parents. Such impressions ultimately contribute to misunderstandings. It is essential that the conference provide a realistic picture of student progress. "Telling it like it is" often allows the parent and teacher to work more closely together in helping the child.

One of the biggest limitations of parent-teacher conferences

---

* The word "all" should be interpreted here to mean "as many as practically possible."

† Student-teacher "conferences" are an excellent means of reporting progress to the student. Generally these informal "conferences" take the form of discussions where the teacher identifies areas of growth and areas where improvement is needed. Such "conferences" may be held any time, and need not be regularly scheduled. They can be used quite effectively to complement the parent-teacher conference.

is the amount of time available for each meeting. Conferences are usually tightly scheduled in 15 to 20 minute intervals. This provides ample time for many conferences; however there are others that require either longer meetings or more efficient use of conference time. Without careful teacher planning, much of the scheduled conference time can pass before substantive issues are discussed. More efficient use of conference time can be made if parents are alerted in advance of their child's progress. An informal note frankly summarizing strengths and weaknesses can be sent to parents prior to the conference. This provides parents time to think about existing problems as well as their possible solution and thereby formulate more relevant questions. Such a pre-meeting notice can contribute substantially to the conference's effectiveness.

Parent-teacher conferences are commonly accepted as an excellent means of exchanging ideas on student progress. Recent years have witnessed a phenomenal growth in the number of schools providing for these conferences. It is generally recognized that this means of reporting student progress is much more meaningful than a formal report card. Certainly this method is more in keeping with the laboratory approach to learning, than is the reporting of a letter grade.

Who should participate in these conferences? Some teachers prefer to meet with students and parents separately, while others invite students to participate. It should be recognized that some children profit considerably from participation in these conferences, whereas others feel very uneasy. The decision as to whether or not to invite students to participate in the conference should be determined with the best interests of the child in mind. Extending a student invitation which is not mandatory in nature is often the best procedure to follow.

Another possibility would be a combination of conferences—perhaps one conference with both student and parent, one teacher—parent, as well as frequent discussions between teacher and student. This arrangement would have many advantages, but realize that it may demand an inordinate amount of teacher time.

Regardless of who attends the conference, the teacher should provide an assessment of student progress that is characterized by comprehensiveness and professional frankness. Data from homework, tests and record forms should certainly be included, as well as a subjective evaluation of student progress in both academic and social areas. Perhaps a check sheet would be helpful in identifying particular areas of concern. All of these data would be synthesized and subsequently analyzed in an effort to provide a valid assessment of student progress.

SELECTED LEARNING ACTIVITIES

1.   Suppose you are going to use one of the Post Numeration System Activities from Chapter 7 for a performance test. Which activity would you use? Why?

2.   Select two of the lessons presented in Chapters 6 and 7. Construct a performance test item for each lesson. (Be sure to identify manipulative materials that may be used.) Administer these performance test questions to a colleague or student who has completed these lessons. What revisions would you make in your questions?

3.   After constructing several performance test items for one of the laboratory lessons (you may use a lesson from this book or one you have developed) field test it with at least three different students. Complete a Record Form Card (Figure 2, p. 236) for each child. Were the same data recorded on these cards? How were these data similar? How do they differ? Would you modify your laboratory lessons as a result of this field testing?

4.   Select a classroom and observe one child for several consecutive days. Suppose you have scheduled a conference with the child's parents.

   a.   Write a pre-conference note to the parents.
   b.   Would you invite this child to the conference? Why or why not?
   c.   Identify specific strengths and weaknesses to be mentioned during this conference.
   d.   Make specific suggestions to the parents of ways in which they might contribute to the students' progress.

BIBLIOGRAPHY—CHAPTER 10

Biggs, Edith E. and James R. MacLean. *Freedom To Learn*. Menlo Park, California: Addison-Wesley, 1969.

Bloom, B. S.; J. T. Hastings; and G. F. Madaus, *Handbook on Formative and Summative Evaluation of Student Learning*. New York: McGraw-Hill Book Co., 1971.

Brownell, William A. et. al. *The Measurement of Understanding*, 45th Yearbook. National Society for the Study of Education, Part I. Chicago: University of Chicago Press, 1946.

Evaluation in Mathematics, 26th Yearbook. Washington, D.C.: National Council of Teachers of Mathematics, 1961.

Kidd, Kenneth P.; Shirley S. Myers; and David M. Cilley. *The Laboratory Approach to Mathematics*. Chicago: Science Research Associates, 1970.

11   Epilogue

Now that you are familiar with the basic concept of the mathematics laboratory and procedures for its implementation, it's time to begin accumulating and developing laboratory experiences for your classroom. The Appendices will provide valuable resource material insofar as the identification of prospective sources is concerned. Keep in mind that these lists may soon be outdated. They will, however, provide a substantial beginning.

Also keep in mind that commercial materials are produced for the mythical "average" student, and therefore may be in need of slight modification for your particular group. Do not be hesitant to make any changes that you feel are necessary. No one knows your students better than you.

As you become more comfortable with the laboratory approach you are likely to become "sensitized" to the infinite number of possible laboratory activities existing in the everyday environment. When ideas occur to you, jot them down immediately, as they are often elusive when an attempt is made to reconstruct them weeks, days, or even hours later. You will find that in a short period of time a "card file" will begin to develop representing unique activities of your own design.

Share ideas with your colleagues and in return they will share theirs with you. Once again we are reminded of the fable:

> If I have a dollar and you have a dollar and we exchange dollars, we both still have a dollar. But if I have an idea and you have an idea and we exchange ideas we both now have two ideas.

Begin slowly. The laboratory approach will be a refreshing change for both yourself and your students, but there is a danger in attempting to completely redefine the basic nature of your mathematics program in one fell swoop. Initially it is wise to confine laboratory experiences to one day per week. This will give everyone concerned an opportunity to adjust to the new classroom atmosphere and instructional setting. At the same time, you can be examining and/or developing other possibilities for future laboratory

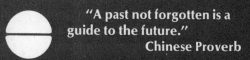

"A past not forgotten is a
guide to the future."
Chinese Proverb

activities. It is very important that your first attempts be somewhat
tightly structured so that students know precisely what is expected of
them. As they (and you) become more comfortable in the laboratory
setting, activities can become more diversified and less structure
need be imposed.

Do not be dismayed if your initial efforts are not an overwhelming success. As indicated earlier, this method of mathematics
instruction places students and teachers alike in a relatively unfamiliar
situation. Initially this unfamiliarity may foster some anxiety. However, students are quick to adapt and are often the first true converts
to a laboratory approach. We are confident that the amount of additional effort required initially will pay large dividends in a short period
of time.

Now it's time to roll up your sleeves and begin. Good luck!

A final thought:

> "You may give them your love but not your thoughts for they
> have their own thoughts. You may house their bodies but not
> their souls for their souls dwell in the house of tomorrow. . . .
> You may strive to be like them, but seek not to make them like you.
> For life goes not backward nor tarries with yesterday."
>
> Kahlil Gibran

APPENDIX A

AN ANNOTATED BIBLIOGRAPHY OF SELECTED MATHEMATICS LABORATORY PUBLICATIONS FOR TEACHERS

This bibliography is prepared to *provide teachers with a core of mathematical ideas and instructional techniques that fit into an active approach to learning mathematics.* We could list hundreds of books, pamphlets, booklets, articles, etc. related to mathematics laboratories. Realizing that most teachers do not have either the resources to procure these materials or the time to read them, this *bibliography has been limited to publications which we feel have the most payoff per dollar spent.** These books have helped us and we feel each of them will be a valuable part of any teacher or school's professional library.

We have purposely avoided pupil oriented materials such as *Exploring Mathematics On Your Own, Experiences in Mathematical Discovery, Experiments In Mathematics, Activities in Mathematics* (AIM) and *Low Achievers in Mathematics Program* (LAMP) materials. Information on these and other similar materials is readily available. In fact, each issue of *The Arithmetic Teacher* and *The Mathematics Teacher* devotes a section to reviewing new materials and products in mathematics for both students and teachers. We hope this list will be useful.

---

* Prices are approximate and are subject to change.

1. Association of Teachers of Mathematics, *Notes On Mathematics In Primary Schools*, New York: Cambridge University Press, 1967.

   A potpourri of lessons for elementary grades and junior high school. No specific sequence is suggested or required. Many interesting mathematical topics are included, although related learning activities and teaching suggestions are scarce. This book is for teachers looking for new ideas as well as for challenging problems to share with their pupils. To profit from this book, you must do some mathematics! (Price—$4.95).

2. Biggs, Edith and MacLean James. *Freedom to Learn*. Menlo Park, California: Addison-Wesley Publishing Co., 1969.

   A book that provides a readable rationale for an active learning approach to learning mathematics. Specific discussion is directed toward the laboratory classroom, how to start and the role of the teacher and principal. Discussion is accompanied by many photographs of students actively learning. Although illustrations are generally directed toward primary grades, this book is an excellent resource material for teachers at all levels. (Price—$7.95).

3. Charbonneau, Mannon P. *Learning to Think In A Math Lab*. Boston: National Association of Independent Schools, 1971.

   An experienced teacher in a mathematics laboratory setting shares some ideas, her remarks are personal, yet practical, and addressed to teachers considering a laboratory approach. The book also provides many tasks or activities related to a variety of mathematical topics. (Price—$2.50).

4. Davis, Robert. *Discovery In Mathematics: A Text For Teachers*. (1964); *Explorations In Mathematics: A Text For Teachers*. (1966) Menlo Park, California: Addison-Wesley Publishing Co.

   These books provide short lessons or activities on a wide variety of topics that can be explored by students independently or in small groups. Although designed for middle grades, most of the ideas can be either introduced or reviewed at other levels. These books can be used in any mathematics classroom and are not uniquely suited to the laboratory. However, the lessons do foster student independence and are generally presented in the spirit of actively involving students in learning mathematics. (Price—$9.00 or less per book).

5. Dienes, Zoltan P. and E. W. Golding. *Exploration Of Space and Practical Measurement* (1966); *Learning Logic, Logical Games* (1966); *Modern Mathematics For Young Children* (1970); *Sets, Numbers, and Powers* (1966).

   These books exemplify an active approach to learning mathematics and provide many challenging ideas for the classroom teacher. Each book provides many different physical embodiments for developing mathematical concepts. All books are available from Webster/ McGraw-Hill N.Y. (Price—$2.45 or less per book)

6. Dumas, Enoch. *Math Activities For Child Envolvement*. Boston: Allyn and Bacon, Inc., 1971.

Describes many activities that can be used to develop mathematics in the elementary school. Most of the activities are geared toward computational skills. Many games using concrete materials are included. This is a resource book for teachers looking for different ways to involve students in learning mathematics. (Price—$4.95).

7. *Experiences In Mathematical Ideas*, (Volumes 1 & 2), Arnold M. Chandler, ed., Washington, D.C.: National Council of Teachers of Mathematics, 1970.

These are two volumes of excellent activities that focus on 13 broad mathematical concepts, including Base and Place Value and Physical Models for Decimals. Each unit provides a series of experiences based on concrete activities. Each experience is organized by pupil objectives, materials, teacher strategy, and activity cards or worksheets. Materials for students can be easily reproduced. The experiences are based on concrete materials and provide different physical embodiments for the mathematical concepts. Although designed for low achievers in grades 5–8, many of the ideas and activities can be used effectively at other levels. These volumes provide valuable resource material for any classroom teacher. (Price—$20.00 per volume).

8. Fitzgerald, William, David Bellamy, Paul Boonstra, John Jones and William Oosse. *Laboratory Manual For Elementary Mathematics. Boston: Prindle, Weber and Schmidt, 1973.*

This book contains mathematics laboratory lessons for teachers that provide small group involvement and utilize concrete materials. Ideas from many of these lessons can be applied directly to the elementary and/or junior high school classroom. (Price—$5.95).

9. Fletcher, T. J. *Some Lessons In Mathematics.* New York: Cambridge University Press, 1964.

This is a companion volume to *Notes on Mathematics in Primary Schools* and assumes a similar degree of teacher independence. Discusses many interesting and challenging mathematical topics. It is designed for mathematics teachers at the junior high level or above. (Price—$3.95).

10. Greenes, Carole E.; Robert E. Willcutt; and Mark A. Spikell. *Problem Solving In the Mathematics Laboratory: How To Do It.* Boston: Prindle, Weber and Schmidt, 1972.

A collection of problems and activities that utilize four different types of concrete materials: Cuisenaire rods, attribute blocks, geoboards and multibase blocks. The lessons are designed for prospective and/or inservice teachers (K-12) in a laboratory setting. Since many of these activities can also be used by teachers with their pupils, the book also provides a valuable source of laboratory problems. (Price—$5.95).

11. *Instructional Aids In Mathematics.* Emil Berger, ed. 34th Yearbook, Washington, D.C.: National Council of Teachers of Mathematics, 1972.

This publication updates the 18th Yearbook, entitled *Multi-Sensory Aids in the Teaching of Mathematics.* It provides much helpful information on manipulatives in mathematics and their use. An excellent bibliography of instructional aids is also included. This Year-book is a must resource for every mathematics laboratory and/or mathematics teacher. (Price—$12.50).

12. Kidd, Kenneth P.; Myers, Shirley; and Cilley, David. *The Laboratory Approach to Mathematics*. Chicago: Science Research Associates, 1970.

A book on laboratory teaching. Provides information helpful in establishing mathematics laboratory activities. Includes an excellent collection of laboratory lessons developing the concept of ratio. Discussions are accompanied by many interesting photographs of pupils in a laboratory setting. Ideas are directed toward upper grades. (Price—$6.25).

13. Laycock, Mary and Gene Watson. *The Fabric of Mathematics*. Hayward, California: Activity Resources Co., 1971.

A reference for elementary and junior high school. Material is organized by topics. Each mathematical topic includes objectives, related manipulative materials, and games, as well as references for children. It contains a good deal of practical information for the classroom teacher. (Price—$12.50).

14. Lorton, Mary Baratta, *Workjobs*, Menlo Park, California: Addison-Wesley Publishing Company, 1972.

This book focuses on the development of language and mathematics skills at the primary level. Topics, such as classification and sets, are developed through a variety of different experiences that actively involve pupils. Each experience is described; needed supportive materials are identified; and teaching suggestions (including ideas for getting started and following up) are presented. Experiences are accompanied by photographs of pupils involved in the activity. This is an excellent resource for preschool and primary grade teachers. (Price—$6.95).

15. Nuffield Mathematics Project Materials:
I Do, and I Understand
Pictorial Representation
Graphs Leading to Algebra
Beginnings (1)
Mathematics Begins (1)
Shape and Size (2)
Computation and Structure (2)
Computation and Structure (3)
Computation and Structure (4)
Desk Calculators
How to Build a Pond
Environmental Geometry
Probability and Statistics
Your Child and Mathematics
Maths with Everything

A uniquely contemporary course of mathematics study for teachers of children aged five to thirteen. These teacher guides suggest many ways to make mathematics more interesting and rely heavily on the children's own drawings and assignments. All guides are available from John Wiley and Sons. (Price—$3.50 or less per book.)

AN ANNOTATED BIBLIOGRAPHY OF SELECTED COMMERCIALLY AVAILABLE MATHEMATICAL GAMES*

INTRODUCTION   To facilitate pupil centered learning activities, several different types of materials have been developed by manufacturers. Included in these materials is the production of mathematical games.

Teachers have found that mathematical games can fulfill many purposes including:

(1)   Making practice periods pleasant and successful

(2)   Teaching mathematical concepts or ideas

(3)   Teaching mathematical vocabulary

(4)   Motivating effective study habits

(5)   Providing for individual differences

(6)   Favorably affecting the attitudes of children toward mathematics

(7)   Improving reading in mathematics

(8)   Providing an enjoyable means of summarizing or reviewing a unit

(9)   Adding to the enjoyment of classwork and homework

It is not proposed here that games represent an appropriate strategy

---

*   The authors are indebted to Ms. Jeanne Bursheim, a teacher in the Willow Lane Elementary School in Osseo, Minnesota, for her permission to reprint some of the material in this appendix.

for the entire mathematics program. However, as one of many useful strategies they can be invaluable to both pupil and teacher.

This bibliography includes: (1) the names of selected mathematical games, (2) suggested grade levels, (3) the name of a manufacturer or distributor, (4) approximate cost, and (5), a brief description. A complete mailing address for each company listed may be found in Appendix D.

The bibliography is limited to information available from the manufacturers and/or distributor. For this reason, some game descriptions are not complete. *Only one manufacturer or distributor is listed for each game, even though many of the games are available from more than one source.* This bibliography is limited to games of chance, skill and endurance played by two or more students, in accordance with a given set of rules. Therefore, mathematical puzzles and other pastimes are excluded. General purpose items which may have games included as part of the child's activity, but which are not intended to be primarily used in this way, are also excluded. A classification system was developed to facilitate the task of locating a particular type of game. It was soon apparent that such a classification system should be made by mathematical topics rather than by grade levels, concepts or purposes. Once these categories were established, the appropriate game descriptions were listed in alphabetical order. The following classification categories were used in this bibliography:

I. LOGIC AND STRATEGY TYPE GAMES

These games are not only limited to games involving logic and strategy, but also to games that have direct mathematical overtones. For this reason, many logic and strategy type games, such as chess and checkers, are excluded.

II. NUMBER SYSTEMS, THEORY, SYMBOLS, AND SENTENCES

Many of these games are designed for early primary-age children and their initial development of number concepts, such as one-to-one correspondence, rote counting, place value, and numeral recognition. Also included are more sophisticated games that involve several mathematical ideas centered around a given theme, such as WFF'N PROOF's *Equations* game.

III. GRAPHING

The purposes of the games included in this category include the utilization of ordered pairs in describing grid locations and the development of corresponding strategies. These games are appropriate from second grade through adult.

IV. GEOMETRY

Most of these games involve the recognition and memorization of shapes and names of regular geometric figures. Other games include ideas such as pattern building, symmetry, and the description of properties of geometric figures.

V. MEASUREMENT

One of the games involves actual length measurement, while another involves questions and calculations concerning various types of measures.

VI. SETS

Ideas such as union and intersection of sets, properties of sets, classification of sets, and comparison of sets are incorporated into the games in this category.

VII. FRACTIONS

A wide variety of games are included. Topics such as equivalent fractions, arithmetic operations on fractions, visual comparisons and relationships between percentages and fractions are considered.

VIII. TIME

The main purpose of these games is to teach children to read a standard clock.

IX. MONEY

These games are designed to help children comprehend the value relationships that existed between different combinations of coins and paper money through actual manipulations of play money.

X. ARITHMETIC OPERATIONS

The majority of commercial games related to arithmetic operations are designed to augment the teaching of addition, subtraction, multiplication, and division of whole numbers. Therefore, to aid the reader who is concerned with a particular operation, this category has been classified into the following sub-categories:
- A. GAMES INCLUDING ADDITION, SUBTRACTION, MULTIPLICATION and DIVISION
- B. ADDITION
- C. SUBTRACTION
- D. ADDITION and SUBTRACTION
- E. MULTIPLICATION
- F. DIVISION
- G. MULTIPLICATION and DIVISION

Most of these games were found to be either the "Bingo" type or card games.

XI. GENERAL ALL-PURPOSE GAMES

As suggested by the title of this category, these games involve combinations of several mathematical concepts. Some of the concepts considered include: inequalities, placeholders, arithmetic operations, equations, set theory, designs, patterns, codes, purchasing, decimals, fractions, percent, geometry, symbols, slide rules, and modular arithmetic.

GAME DESCRIPTIONS

I. LOGIC AND STRATEGY TYPE GAMES

Haar Hoolin Perception Games (grades 1–12, Selective Education Equipment–SEE, Inc., No. ALP001, price $3.50) The purpose of this game is to increase the children's capacity to visualize abstract concepts. The game requires the player to visualize the cards in varying combinations in order to form patterns. The set includes a series of 15 games, 3 forms of solitaire, 10 strategy games for 2 to 4 players, an ESP game for 2 to 6 players, and a party game. The games are graded according to difficulty.

Kalah (grades 2– 12+, Math Media Inc., No. M310, price $3.00) Kalah is an ancient African game which uses the fundamental processes of mathematics. The element of chance is absent, which means the more skillful player will be victorious. The board is made of simulated walnut. Instructions are included.

MEM (grades 5–12, Selective Education Equipment – SEE, Inc., price $6.50) The purpose of this game is to stimulate the ability to evaluate and perform strategic alternatives. The game is played with 32 colored stones. The basic game idea is to build stone patterns on the table which represent patterns remaining on the board. Two different instruction books are included which provide for four different games including MEM solitaire.

Snag (grades 3– 12+, Math Media Inc., No. M304, price $5.50) The object of this game is to accumulate the most tiles, using a combination of alertness and memory. The game consists of 76 numbered tiles, 4 blank replacement tiles, 4 racks, 1 playing board, 1 score pad, 2 pencils, and an instruction book. Two, three, or four players can play.

3-D Tic-Tac-Toe (grades K– 12+, Selective Education Equipment– SEE, Inc., No. 444444, price $4.00) The object of the game is to place 4 pegs in a straight line, either horizontally, vertically, diagonally on the horizontal plane, or diagonally on the vertical plane. There are 76 possible ways of winning. The six inch transparent square playing surfaces and the design of the structure lend themselves to easy "sighting." It can be played by 2 to 4 players. Instructions are included.

Wff (grades 3–6, Wff'N Proof, price $1.25) This game provides practice in constructing and recognizing WFF's (Well Formed Formulas—expressions in mathematical logic).

Wff'N Proof (grades 1– 12+, Wff'N Proof, price $6.00) This game provides practice in abstract thinking while providing an opportunity to learn mathematical logic. It consists of a 21 game kit that starts with games that can be mastered by six-year olds and ends with games that will challenge intelligent adults.

Who Is Your Favorite? (grades 3–6, Stanley Bowman Co., Inc., No. 113–3M, class—pack of 15 units cost $15.50) This is a unique "computer" type game that permits the child to become involved with solving a problem—in this case—identifying one of 32 boys and girls. This unit incorporates binary functions. In the process of playing, the student quickly learns the logic of the game and the validity in the binary set approach. The game can be played by an entire class or by small groups.

II. NUMBER SYSTEMS, THEORY, SYMBOLS, AND SENTENCES

Abacus Spinner Game (grades 2–6, Creative Publications, price— $2.95) A game designed to provide practice in recognizing and understanding place value. The tic-tac-toe format is used by students to locate the numeral representation of the abacus on the spinner card.

Equations grades 4–12+, Wff'N Proof, price: kit $5.00, teacher's manual $1.00) The object of the game is for each player to attempt to build his own equation, while preventing his opponents from completing theirs. The first player rolls mathematical cubes out on a table and stipulates a common "Goal" of an equation, e.g., the right side of the equals sign. For a given roll of the cubes, this Goal is not changed after it is set. The rest of the game is aimed toward shaping the left side of the equation, which is called the solution. By moving just one cube at a time, each succeeding player has the opportunity to forbid, require, or permit, in building the solution, any of the numerals or operation signs that face upward on the cubes. Any accurate means of reaching the Goal of the equation is acceptable, provided the numerals and operation signs necessary are available on the upward faces of these cubes. The instruction manual for the kit explains advanced forms of the game, permitting the introduction of such ideas as binary, octal, arithmetic, modular arithmetic, logarithms, and complex numbers. The game can be as easy or difficult as the players wish. The game comes in a plastic box which contains 32 cubes imprinted with numerals, arithmetic operation signs, a playing mat, a timer, and an instruction manual. The game can be played by 2 to 4 players.

First Arithmetic Game (grades K-1, Garrard Publishing Co., Dolch, price $2.00) This game is designed to help interest children in counting. It includes a set of large cards each containing a four square by five square array. The first row consists of a set of numerals or number words. The child then places a small card with a picture of a set of similar objects under the appropriate numeral.

Heads Up (grades 2–12+, Math Media Inc., No. M302, price $5.00) The purpose of this game is to earn the highest score by building four basic equations from 14 numbered cubes before the time runs out. The game includes a three-dimensional playing field, a cup, 14 cubes, and a timer. Any number of players may play.

Krypto (grades 4–12, Creative Publications, price $1.25) Pupils combine cards from a Krypto deck to construct object numbers. It provides drill in basic operations and is appropriate for a wide range of abilities. The game may be played by one or several children.

Ladder Race (grades K-1, Childcraft Education Corp., No. OX269, price $4.50) The purpose of this game is to help children learn basic number concepts. Two players race the two wooden figures up a 20 inch plywood ladder. Numbered tickets, picked at random by the players, decide how many steps in each move.

Math Bingo (grades K-1, Nifty Division, St. Regis Paper Co., No. 9590, price $5.50) The game includes 40 student game cards, 400 card markers, 30 call out cards (1–10 in △, ○, □), 3 master cards, and a teacher's guide. The game is designed to help children learn the numerals 1–10 and the names and shapes of three geometric figures.

Number Group Recognition Game (grades K-1, Childcraft Education Corp., No. OX184, price $1.25) This game introduces counting and number group recognition. The "key" cards tell players which kinds of cards they are to collect. The first player to collect 6 cards of the same value group is the winner. The game is played by 2–4 players.

Numeral Dominoes (grades K-1, Childcraft Education Corp., No. OX264, price $5.00) The purpose of this game is to reinforce recognition of the numerals 0 to 9. The brightly colored super-size dominoes feature numerals 0 to 9 and can be played by standard domino rules. The game includes a fabric bag.

One, Two, Three, Think! (grades K-3, Selective Education Equipment—SEE, Inc., No. XGAMO4, price $2.25) The unit consists of 36 cards and instructions which are to be used in various attribute card games. The attributes included are size, shape, color and fullness.

Pacemaker Games Program (grades K-1, Fearon Publishers, price $48.00 per set) The purposes of this program as stated by the manufacturer are: mastery of general game skills; ability to count by rote to 10 or higher; ability to enumerate 8 or more objects; ability to use ordinal numbers to 5 or more; one-to-one correspondence; recognizing small groups to four or more; recognizing common shapes; recognizing colors; identifying coins; and using quantitative vocabulary to recognize numbers. The game is described as a pre-arithmetic readiness program that includes table search games and active racing games. The materials include: a 72 page teacher's manual (instructions for 65 games graduated in difficulty), 5 game boards, 5 different sets of playing cards, 2 pair of dice, a set of dominoes, a whistle, counters in 8 geometric shapes, and 9 common colors.

Place Value I (grades 2–6, Creative Publications, price—$2.95) Self-correcting spinner card provides practice in recognizing place value through 1,000. Other activities include using the greater than, less than, and equivalent symbols.

Place Value II (grades 4–8, Creative Publication, price—$2.95) Students read numbers through the hundred millions. It includes activities in Place Value I plus an addition drill in the scoring procedure.

Prime Drag (grades 6–10, Creative Publications, price—$3.00) Game objective is to have students identify primes and composites. Chance cards add to fun and excitement as well as testing student's knowledge of multiples. Two to four players.

Prime Factor (grades 6–9, Creative Publications, price—$4.50) A game which requires the players to factor in order to move, purchase properties or pay rent on properties owned by another player. Colors emphasize the different sets of numbers providing a strong visual image.

Prime-O (grades 5–9, Creative Publications, price—$1.95) An exciting card game which emphasizes prime numbers, composite numbers and their relationships. Provides thorough practice in prime factorization.

Rack-o (grades 5–12, Milton Bradley Co., No. 4765, price $3.50) This game includes a deck of cards and a specially designed card rack. The object of the game is to strategically place the cards in numerical sequence. The cards are inserted in colorful plastic racks.

Rank-o (grades 5+, Midwest Publications Company Inc., price $8.95) Four different games employing drill and strategy in ordering cards. The complexity of game can be controlled by the decks of cards used. The game can be played in groups of two to four.

The Real Numbers Game (grades 3– 12+, Wff'N Proof, price $1.00 or deluxe $1.50) This five game kit is designed to prepare beginners for playing "Equations." The equipment comes clipped to a ball-point pen, handy for play anytime that two or three minutes is available. It deals with real, rational, irrational, integer, and natural numbers.

Symbol (grade level 2–6, Stanley Bowman Co., Inc., No. 113–IM, 113–1T, 113–1MT, complete kit cost $39.50) This game is played like "Bingo" with 30 different boards plus a set of "callers." The boards are marked with mathematical symbols while the calling pieces are flash cards of the actual symbols. Each game board is different, so there is a winner for each game. The game can be played by the entire class or by small groups.

The Ten Game (grades 1–2, Garrard Publishing Company—Dolch, price $1.25) The game is designed to teach the naming of numbers and how to put numbers together, all within the frame of ten. The game consists of a deck of numeral cards with each card having a given numeral, the number word, and a representative array.

III. GRAPHING

Battleship Strategy Game (grades 2– 12+, Miles Kimball Co., No. 3834, price $4.95) In this game each player or team has his own fleet of plastic ships that range from an aircraft carrier to a submarine. The object of the game is to sink the opponent's ships by firing salvos or shots and calling out the strike area, without being able to see where his ships are positioned. The purpose of the game is to utilize ordered pairs to describe a location on a grid.

Vectors (grades 4–8, Selective Educational Equipment, Inc.— SEE, No. IN1128, price $5.00) The purpose of this game is to provide experience in the fundamentals of graphing using the x and y axis. As the players progress they are given the freedom to make up harder rules to provide a more strategic play. It can can be played by two to four players. One game consists of a pegboard, 20 pegs, and one double pack of cards marked in numerals with + and − signs.

IV. GEOMETRY

Beeline (grades K–12+, Selective Educational Equipment, Inc., SEE, No. BEELIN, price $4.25) "Beeline" utilizes logic and strategy while attempting to make a complete beeline across the playing board. Each hexagonally shaped playing piece has either a straight line, obtuse angle, or acute angle printed on it. These playing pieces are set in front of the player in four identical universes before he can use the pieces in the next universe. A fourth type of playing piece, "the node," is used as a multi-directional piece and can be played either on the player's own beeline or on his opponent's beeline. This game can also be used to play "hex." It consists of a playing board, 84 playing pieces and instructions.

Configurations Game (grades 4–12+, Miles Kimball Co., No. 4429, price $5.00) This game is based on the book "Geometry of Incidence." The kit includes 2 puzzle boards, 15 geometric figures, and 50 reverse-draft numerals—all stored in a plastic box.

Euclid (grades 8– 12+, Midwest Publication Company, Inc., price $6.50) A card game of rummy that uses principles of plane Euclidean geometry. Develops skills in the use of axioms and logic in proofs. It can be played in small groups without teacher direction.

Geometric Dominoes (grades K-4, Selective Education Equipment, Inc. SEE, No. GN1159, price $1.35) The purpose of this game is to teach quick recognition and identification of geometric shapes. The game also provides training in matching shapes and patterns. The game consists of 24 domino cards in a box with instructions. It can be played by 1 or 2 players or by 2 teams.

Geometric Matching (grades 4–9, Selective Education Equipment, Inc., SEE, No. ESA005, price $1.35) This game is to be used after geometrical terms have been introduced. There are 52 cards, some of which are printed with diagrams of geometrical properties, while the remaining cards have written descriptions of each property (e.g., "alternate angles are equal," "equilateral triangle"). Complete instructions are included.

Geometric Shapes Spotting (grades K-2, Selective Educational Equipment, Inc. -SEE, No. GN1046, price $2.00) The purposes as stated by the manufacturer are to encourage players to be aware of shape and form, to stimulate an interest in mathematical properties and to develop sensory powers. The game is played like bingo with the shapes taking the place of numerals. It can be played by 2 to 4 players and includes 4 playing boards and 32 cards.

Polygons (grades 6–12, Creative Publications, price $1.95) Players form "books" of four cards each related to the same geometric figure. Students will become thoroughly acquainted with the most important properties of the figures as they play the game.

Psyche-Paths (grades 4 and up, Cuisenaire Company, 20080, price $2.95) A puzzle game that involves topological notions. Children construct colored paths using hexagonal pieces in an effort to reach their destination before their opponent. This game can be played by one or several children.

Shape Analysis Matching Game (grades K-3, Responsive Environment, Corp. -REC, Learning Materials Division, No. 105-G-108, price $3.00) The purposes of this game are to develop geometric shape recognition and to pose problems in observation and discrimination. Each of 2 games includes 18 pairs of cards. One card of each pair has geometric shapes in random arrangement while the other card has the same shapes in an ordered array. The child pairs the matching cards.

Shapes Lotto (grades K-1, Playskool Manufacturing Co., No. 599, price $15.00) The players of this game learn to recognize various geometric shapes. The players match wooden tiles in 6 geometric shapes to the same shapes on the lotto boards. As the children progress they must place the shapes on a lotto board wherein the shapes are embedded. The set consists of 8 lotto boards, spinner board, 72 wood tiles in six shapes, 2 sets of cue cards, 4 bags and an instruction book.

Symmetry Dominoes (grades K-3, Selective Educational Equipment, Inc.—SEE, No. ARN656, price $2.75) The 28 symmetry dominoes are 2″ × 3″ plywood, the ends of which are designed to create symbols of simple recognizable shapes when matched against another domino. There are five series of designs: operational symbols ($+$, $-$, $\times$, $\div$, $=$), symmetrical capital letters, geometrical shapes, toys, and flowers.

V. MEASUREMENT

Inch By Inch (grades 1–3, Selective Education Equipment, Inc.—SEE, No. GNINCH, price $3.75) This game simplifies the concept of measurement through practical application.   The children follow instructions on the cards which provide exercises in measuring distances on a colored board. The game includes 12 "lucky bag" cards, large die, 6 moving pieces, 6 graduated 1 foot cardboard rulers, and complete instructions. The game can be played by 4 to 6 players.

Liquid Measure Game (grades 3–8, Creative Publications price—$2.95) The spinner card illustrates liquid measures in attractive color. Students find the equivalent measurements on activity cards. Self-correcting.

Spin-A-Gallon (grades 2–8, Creative Publications, price—$2.95) Game provides drill in forming equivalent liquid measurements. Includes four-color spinner card illustrated measurement cards and self-correcting checksheet.

Which Is More? (grades 2–4, Selective Education Equipment, Inc.—SEE, No. GNO549, price $2.00) In this game, for 1–5 players, are questions on time, capacity, and measure. One player controls the cards and the others have to calculate the answers. The correct answers are on the backs of the cards. Typical questions are: "Which is more? 2 feet or 26 inches?" The game includes 36 heavy cardboard cards.

VI. SETS

On-Sets (grades 3–12+, Wff'N Proof, price $4.00) The kit includes 30 games on set theory. The basic operations of unions and intersection of sets as well as inclusion and identity are presented in a fast-moving game.

Spot the Set (grades K-4, Selective Education Equipment, Inc.—SEE, No. GN1158, price $2.00) The purposes of this game are to discover the properties of collections through the classification and comparison of shapes and sets, and to help children see and relate order and pattern. It is a "bingo" type game with 4 base boards and 32 small cards with sets of geometric shapes. It can be played by 1 to 4 players.

VII. FRACTIONS

Action Fraction (grades 3–6, Schoolhouse Visuals Inc., No. M320, M321, M322, price $7.00 each) The purpose of "Action Fractions" is to develop concepts and skills involving fractional numbers. There are three games—Circles (M320), Squares (M321), and Triangles (M322). In each five different skill level games can be played by up to eight players. Each player rolls the fraction-marked cubes to determine what fractional part he will place or remove from his gameboard. As the action progresses many combinations occur that can be replaced by fractional parts of equivalent value. Players learn to exchange two fractional parts for a single part (e.g., $\frac{1}{4}$ and $\frac{1}{12}$ for $\frac{1}{3}$). Students learn addition and subtraction, renaming fractional numbers, mental computation, problem solving, and visual comparison of fractional parts. Each of the three games contain 8 gameboards, 72 fractional parts, 2 sets of fraction marked cubes, and instructions.

Come Out Even (grades 4–8, Holt, Rinehart and Winston, Inc., price $1.25) This game teaches mastery of addition of fractions by finding least common denominators and equivalent fractions. The game includes two decks of cards, each containing 52 cards. Deck A shows halves, fourths, eighths, and sixteenths. Deck B shows halves, thirds, sixths, ninths, and twelfths. The play is similar to rummy and is designed for 2 to 6 players.

Competitive Fractions (grades K-4, Selective Education Equipment, Inc.—SEE, No. GN0439, price $3.25) This game reinforces the children's ability to understand and mentally add the simple fractions of $\frac{1}{4}$, $\frac{1}{2}$, and $\frac{3}{4}$. The playing pieces are based on a whole square, 4″ × 4″, which is cut into quarter or half segments in different combinations. Moves of the players are dictated by the "activity cards," (e.g., "Pick up $1\frac{1}{2}$.") The players have a choice of which pieces they will take from the umpire's board. The game consists of a box to hold the fractional pieces, playing cards, 3 16″ × 16″ playing boards, and instructions.

Fractions (grades: Set 1, 8–12+, Set 2, 2–8, Gel-Sten Supply Co., Inc., price $1.50) This game consists of a deck of cards that are used to reinforce many concepts involving fractions.

Fraction Dominoes (grades 2–6, Selective Educational Equipment, Inc.—SEE, No. ARN612, price $2.75) The purpose of this game is to reinforce understanding in the direct relationship among the numbered, the named, and the pictured fractions. For example, a circle showing a $\frac{1}{3}$ pie section may be matched with either a domino which reads "$\frac{1}{3}$" or one which reads "one third." The set contains 28 wood dominoes.

A Game of Fractions (grades 5–9, IMOUT Arithmetic Drill Games, price $6.30) This game consists of 48 cards, 8 of which are for special drill purposes and are of proper and improper fractions. The game is played like "Bingo."

Make One (grades 3–6, Garrard Publishing Company—Dolch, price $1.50) The purpose of this game is to teach understanding of common fractions and percentages. The game is played by adding fractions or percentages that combine to make "one."

One (grades 6–12, Creative Publications, price—$1.85) A variety of challenging games in which players form an object number by combining the whole numbers and fractions on their cards. Provides practice with the basic operations.

Recognizing Fractional Parts (grades 4–8, Creative Publications, price—$2.95) A game with a tic-tac-toe format which provides practice in recognizing fractional parts. Includes game activities in addition and subtraction of fractional numbers and self-correcting student checksheet.

VIII. TIME

Tell Time Quizmo (grades K-2, Milton Bradley Company, price $2.50) This game is designed to teach children to tell time. It is played like "Bingo" and consists of a large clock dial with metal hands for the instructor's use and 39 answer cards with clock dials and movable hands for the students.

"Time-Please" (grades K-1, Gel-Sten Supply Co., Inc., price $1.25) The purpose of the game is to teach children to read a clock face. The game is played by the teacher and a child.

IX. MONEY

Count Your Change Game (grades K-2, Milton Bradley Company, No. 7635, price $3.00) Designed to expose children to the various coin combinations that equal one dollar, this game consists of a gameboard and play money. It can be played by 2 to 8 players.

Pay The Cashier (grades 1–2, Garrard Publishing Co.—Dolch, price $4.50) Each player is given metal coins and paper play money that is the same size as real money. With this money children buy all kinds of things at the Dime Store, the Department Store, the Super Market, etc. During the game each child has a chance to be cashier.

Spin-A-Coin (grades 2–8, Creative Publications, price—$2.95) Lower grade students determine money values shown on the spinner card; upper grade activities include addition and subtraction.

X. ARITHMETIC OPERATIONS

A. GAMES THAT CONTAIN ADDITION, SUBTRACTION, MULTIPLICATION AND DIVISION

Arithmetic Can Be Fun (grades 3–8, Mainco School Supply Company, No. 499145–130/2 Multiplication and Division, price each set $1.25) Each game consists of two decks of playing cards and 6 playing boards. Its purpose is to teach the basic fundamentals. Up to six can play.

Block It! (grades 4–12, Creative Publications, price—$2.95) Players search through the possible addition and multiplication combinations with three numbers from one to six to form an answer that will place three markers in a row to win.

Imma Whiz (grades 3–6, Kenworthy Educational Service, No. 2218 Addition and Subtraction, No. 2219 Multiplication and Division price is $1.50 per game) This game is designed to provide reinforcement with the basic facts. It's a "Bingo" type game that can be played by 2 to 24 players.

"I Win" (grades K-8, Gel-Sten Supply Company, Inc., price per deck is $.98) "I Win" is a set of 18 different card games for children of all age levels. The object of each game is to match the black problem cards with the red answer cards. Two or more can play any game. Each deck includes 25 problem cards and 25 answer cards.

Math Magic (grades 2–6, Cadaco Inc., No. 285, price $6.00) A box containing 4 Bingo cards, a dual spinner card, 3 dice, a shaker cup, 200 markers, a sand timer, a "81" gameboard, and a "216" gameboard. It features five different games which are designed to teach addition, subtraction, multiplication, and division. It can be played by 2 to 4 players.

Number Rummy (grades 1–3, Kenworthy Educational Service, No. 2023, price $1.50) The game consists of a deck of cards whose purpose is to teach concepts, facts, and relationships of numbers.

Numble (grades 4–12+, Math Media Inc., No. M306, price $4.00) This is an addition, subtraction, multiplication, and division game which consists of numbered tiles from 0 through 9 that are to be layed out in ascending or descending order to make totals divisible by three. Special spaces on the gameboard affect the point totals. A chart printed on a board makes the figuring easy for rapid play. The game also contains 4 finished racks, 87 tiles, full directions and a storage box.

Operations Bingo (grades 2–7, Creative Publications, price —$7.50) A bingo game designed for use with the entire class. Tests students on their knowledge of fundamental operations. Twelve separate ditto master sheets enclosed enables the teacher to select the operation and level of difficulty. Each student selects his own bingo card on the dittoed matrix.

Quizmo (grades 1–5, Milton Bradley Co., No. 9309 Addition and Subtraction, No. 9310 Multiplication and Division, price $2.00 per game) The game is played like "Bingo" and consists of cards for the entire class plus teacher call cards and markers. The teacher calls out combinations while the children put markers on the correct answer.

Say-it Games (grades 1–6, Garrard Publishing Company—Dolch, each game—Addition, Subtraction, Multiplication, and Division—is $1.98) The Say-it Games teach the fundamental combinations and is played like "Bingo." The combinations are arranged on the cards in each set from the easiest to the hardest. Teacher direction is not necessary.

Stocks and Bonds (grades 5–12+, a 3M game available through SEE, No. MMB 170, price $7.95). The game consists of stockboard, chalk, blackboard eraser, die, securities, review cards, record sheets, situation cards, stock certificates, and an instructor's folder. The students continually create their own problems in addition, subtraction, multiplication and division. It can be played by any number of people.

Triscore (grades 3–12, Creative Publications, price—$2.95) A variety of novel games which provide practice in addition, subtraction, multiplication, division, and combinations of operations.

Twin-Choice (grades 1–6, Holt, Rinehart and Winston, Inc., price $1.25 per deck) This game consists of 8 separate decks of cards each containing 52 cards. Computation on the cards increases in difficulty from deck 1 to deck 8. The purpose of the games is to teach addition, subtraction, multiplication and division of whole numbers.

B. ADDITION

Addo Arithmetic Game (grades 1–5, Kenworthy Educational Service, No. 2220, price $1.65) The purpose of this "Bingo" type game is to study, drill, and review the one hundred addition combinations. Thirty-six players can play at once.

Add-Me (grades 1–4, Gel-Sten Supply Co., price $1.75) This game teaches the 64 basic addition combinations with sums through ten. Contains 2 decks of cards—a deck of 32 addition cards with addition combinations, and 32 cards with corresponding answers.

Big Zero (grades 1–4, Creative Publications, price—$1.85) A card game used to develop meaning in addition. It can also be used as flash cards, group response cards or individual practice cards.

Cover-Up (grades 2–6, Selective Educational Equipment—SEE, Inc., No. COVRUP, price $2.25) The purposes of this game are to provide practice in addition, analyses of various combinations of addends, and an awareness of probability. One or more players can play this game, which involves throwing a pair of dice and covering only combinations of numerals (on the playing board) whose values total those shown on both dice. Each player repeatedly throws the dice until he no longer can cover a numeral. The total value of the remaining numerals represents the player's score. The lowest score wins. The game consists of a plastic playing board, dice, and instructions.

Pyramid (grades 1–12+, Selective Educational Equipment—SEE, Inc., No. PYRMID, price $2.25) "Pyramid" combines vocabulary, spelling, and some simple addition into one game. Fifteen alphabet dice are thrown, the 3-minute timer is started, and the player builds as many words—crossword style—as he can on the numbered grid. In scoring the student adds by 5's. He adds the value of the numerals covered on the grid and subtracts 5 points for each unused die. The game consists of a gameboard, 15 die, a 3-minute timer, and a box of instructions for three game variations.

Tally (grades 1–6, Creative Publications, price—$2.25) One to four students play this game in which number sentences using the addition facts are formed.

Three-Dimensional Dominoes (grades 2–12+, Miles Kimball Co., No. 4799, price $2.19) This game has 45 plastic pieces in triangular shape. Each has three numbered sections to be matched as in conventional dominoes. There are six new domino charts and one new game of solitaire dominoes. Full directions and charts are included.

24-Dice Game (grades K-3, Selective Educational Equipment—SEE, Inc., No. THRADD, price $1.25) This is a dice game where the values shown on the dice are added. It consists of 5 large die, 48 counting chips and instructions.

Yahtzee (grades 3–12+, Math Media Inc., No. M303, price $2.50) The game consists of a set of dice, a shaker, a score pad and chips. It reinforces basic addition skills and can be played by 2 or more people.

C. SUBTRACTION

Subtract-Me (grades 1–4, Gel-Sten Supply Co., Inc., price $1.75) Teaches the 64 basic subtraction combinations with minuends of 10 or less. It contains two decks of cards. A deck of 32 subtraction combination cards and 32 corresponding answer cards.

"Sum" Difference (grades 1–4, Creative Publications, price—$1.95) An excellent companion to "Big Zero", this card game gives practice in subtraction and helps clarify its meaning.

D. ADDITION AND SUBTRACTION

A Game of Addition and Subtraction (grades 2–3, IMOUT Arithmetic Drill Games, price $6.30) This "Bingo" type game consists of 20 addition cards, 20 subtraction cards, 8 combination addition and subtraction cards, and a spinning device.

Making Tens (grades K–1, Childcraft Education Corp., No. OX 175, price $2.50) The purpose of this game is to teach addition and subtraction to ten. It consists of 37 cards and rules, and can be played by 2 to 6 players.

Smarty and Sum Fun (grades 2–8, and K-3 respectively, Ideal School Supply Co., No. 7703, price $3.50 each) This "Bingo" type game is designed to teach the addition and subtraction facts. It can be played by 2 to 6 players.

Teach Key Math (grades K-2, G. W. School Supply Specialists, No. ME-1200, price $6.95). This game includes the Teach Key Schoolhouse, with number pocket, 2 keys, 48 math cards, and instructions. Illustrated cards are "unlocked" when the player finds the missing number in the problem.

E.   MULTIPLICATION

Multifax (grades 4–9, Creative Publications, price—$2.25) A game similar to tally, except that number sentences formed are based on the multiplication facts.

Multiplication: Orbiting the Earth (grades 3–8, Scott Foresman and Co., No. 02157–69, price $3.75). The students move to each of nine space stations orbiting the earth by correctly multiplying the numbers on tossed cubes. The space stations list products of 3, 4, 6, 7, 8, and 9. If the station does not list the product of the numbers he has tossed, the student tries again on his next turn. The game includes a vinyl playing field, 2 wooden cubes, 6 disc markers, and instructions. It can be played by 2 to 6 players or teams.

Multo (grades 3–6, Kenworth Education Service, No. 2221, price $1.65). The purpose of this "Bingo" type game is to drill and review the 100 multiplication combinations. It contains 100 blank deposit coins, 100 calling coins, 36 game cards, a master banking sheet, and instructions.

Sum-Times (grades 1–6, a 3-M game available from SEE, No. MMS150, price $6.95). "Sum-Times" is a mathematical crossword game for 2 to 4 players who try to score the most points by playing up to four tiles in a row. The row of numerals on the tiles must add up to the "key number" or a multiple of it. The thought process involves addition, multiplication and division. The relationship of multiple addition to multiplication becomes apparent to the students as they play the game. The game includes 8 plastic "key-numbers" (2–9), 90 playing tiles with numerals, a playing board, 8 reusable scoreboards, and 4 crayons.

The Winning Touch (grades 3–10, Ideal School Supply Co., No. 7702, price $4.50). The purpose of this game is to reinforce the multiplication facts through "12 times 12." The game is played on a multiplication matrix and the players have numeral pieces on a wood tray—similar to "Scrabble". Designed for 1 to 4 players.

F. DIVISION

Division: Orbiting the Earth (grades 3–8, Scott Foresman
and Co., No. 02157–69, price $5.04). By using his knowledge
of division with quotients through 9, the student moves to
each of seven space stations orbiting the earth. He tosses 3
numeral cubes—any one cube may be the divisor, the other
cubes form the dividend. By arranging the cubes to result in
a quotient listed on the space stations, the student moves
easily from start to finish. The game includes a vinyl playing
field, 3 wooden cubes, 36 remainder cards, 6 disc markers,
and instructions. The game can be played by 2–6 individual
players or teams. Similar "orbiting the earth" games are
available for addition and subtraction.

Rally with Remainders (grades 4–9, Creative Publications,
price—$2.50) A game which provides interesting division
practice at four levels. Self-correcting checksheet included.

G. MULTIPLICATION and DIVISION

A Game of Multiplication and Division (grades 4–5, IMOUT
Arithmetic Frill Games, price $6.30). This "Bingo" type game
includes 40 multiplication and division cards (tables 6–9)
and 8 multiplication and division cards—tables 10–12.

Quinto (grades 2–12+, a 3-M game available from SEE, No.
MMQ160, price $7.95). Two to four players play on a $11\frac{1}{4}$" ×
16" plastic tile board which contains 216 tile-holder squares.
It is based on multiples of 5 (younger students may base
their games on multiples of 2, 3, or 4). The quinto, played in
crossword fashion, provides students with enjoyment, while
they learn addition, multiplication and division. It consists of
95 tiles, a playing board, a grease pencil, a reusable score-
board, and instructions.

XI. **GENERAL ALL-PURPOSE GAMES**

Arithmecubes (grades K-6, Scott Foresman and Company, No.
02156–68, price $3.45). Using 16 one-inch cubes, students can
play 24 mathematical games, ranging from matching sets to
constructing complex mathematical sentences. Twelve cubes
show numerals, two show operation signs, one shows signs for
equalities and inequalities, one shows placeholders. Instructions are included.

Chrominoes (grades K-6, Products of the Behavioral Sciences,
No. 114–1 Chrominoes-Colors, No. 114–2 Chrominoes-Designs,
No. 114–3 Chrominoes-Numerals, No. 114–4 Chrominoes-Color
Coded Dominoe Dots, price $6.00). This flexible game can be
used as building blocks, mosaic designs, or for the regular
domino games of BLOCK, MATADOR, and FIVES. The game
can be played with colors, designs, numerals or spots.

Flinch (grades 1–10, G. W. School Supply, No. 620–P693, price $3.00). This game utilizes a deck of 150 cards made of 10 suits or series, each numbered from one through 15. Rules for several versions are included, with a special children's rules for a game called GO TO THE STORE. A sturdy plastic rack is included.

Games AGDP (grades 1–6, Motivational Research, Inc.). These games consist of 10 different game boards incorporating races using autos, airplanes, Indians, pirates, and par 4 golf. The purposes of these games as stated by the manufacturer are: to motivate students and reinforce skills at 10 levels of difficulty in 17 math areas: addition, subtraction, multiplication, division, symbols, whole numbers, decimals, fractions, substitutions, percent, equations and geometry. Placement tests are used to determine strengths and weaknesses. Four children compete in each game at their own level.

Operational System Games (grades 3–12, Webster Division, McGraw-Hill Book Co., No. 611301, price $12.00). A large number of games can be played with these materials which consist of 8 game surfaces, 120 playing pieces, 4 table cards, and two circular slide rules. The games suggested in the manual (26 basic games) are divided into five categories: clock arithmetic, modular arithmetic, sets, common divisors and multiples, and modular games in the plane and space. These games may be played by 2, 3 or 4 students or teams.

Rook (grades 2–12 + , G. W. School Supply, No. 619–P700, price $1.50). This 4 suit numerical deck of cards can be used to play 23 different games. Rules are included.

TUF (grades 1–12 + , Creative Publications, price $8.00). TUF is a series of games based on equations. Initially they are simple addition and subtraction equations. The games progressively advance by adding more mathematical operations and concepts such as: multiplication, division, parentheses, fractions, decimals, negative numbers, other bases, ratio, proportion and percentage. Later games become more complex with exponential powers, and fractional roots both positive and negative, logarithms, tangent, pi, *i*, and an unknown *x*. The rules of the games are simple enough to master in a few minutes. On a given signal 2, 3, or 4 players each throw their cubes and work simultaneously and individually to construct the longest possible equation. They compete against each other and time. The play is continuous, so the players do not have to wait turns. One set has 60 cubes, 4 blank cubes as extras, 3 timers, 1 rule book and a container. The game can also be played by a single person.

APPENDIX C

ANNOTATED LIST OF SELECTED ACTIVITY, TASK OR ASSIGNMENT CARDS*

Each listing uses a similar format and provides the following information:

a. Name of material.
b. Brief description.
c. Availability of teacher's guide.
d. Suggested grade levels.
e. Name of distributor (Complete address can be found in Appendix D. Only one distributor is listed, even though several may handle the material.).
f. Catalog number, if any.
g. Current price.

ACTIVITIES COORDINATED WITH SPECIFIC PHYSICAL MATERIALS

Geocards—144 cards to be used with a geoboard. Ideas and suggestions for exploring triangles, quadrilaterals, angles, symmetry, area, perimeter, similarity and transformations via the geoboard are pre-

Cuisenaire Co. of America, Inc. Cat.#20040 $15.00

---

* This list of commercially prepared software is not exhaustive, but is representative of the wide range of materials currently available. All of the materials are independent of a particular mathematics program and could be easily integrated into most classrooms. These materials are designed to "involve pupils in doing mathematics" and generally follow the software format discussed in Chapter 5.

sented. Appropriate topic is identified on each card. Cards provide for individual and/or small group exploration. Notes to teacher included. (Grades 3–8).

Geosquare Activity Cards—120 cards designed for use with a geoboard. Cards include developmental activities, games and exercises for individual pupil exploration. (Grades 3–6).	Scott Scientific Cat.#4–662	$9.95

Logic and Sets Work Cards—72 cards divided into two stages. Stage A consists of 36 cards that develop the universal set, empty set, subset and set complement together with properties of these sets. Stage B also consists of 36 cards and develops set operations (Grades 3–9).	Educational Supply Association Limited	
	Stage A Cat. # 9890/920	$4.75
	Stage B Cat. # 9890/939	$4.75

Math Mates—8 sets of 10 activity cards, each designed for use with different manipulative materials. Each card identifies an overall objective. Students may work alone or in small groups. Reading is necessary; however some activities could be used with nonreading students if oral directions are given. Teacher's guide available (Grades 1–8).	Learning Research Associates, Inc.	
	Cat.# m4100 Math Balance	$4.00
	Cat.# m4103 Abacus Board	$4.00
	Cat. # m4133 100 Number Board	$4.00
	Cat.# m4141 Geo-Strips	$4.00
	Cat.# m4143 Mosaic Shapes	$4.00
	Cat.# m4250 Clinometer	$4.00
	Cat.# m4354 Primary Shapes	$4.00
	Cat.# m41107 Attribute Blocks	$4.00

Mathematical Balance Work Cards—20 cards that are coordinated with a mathematical balance or balance beam. Cards present tasks that develop both concepts and skills in addition, subtraction, multiplication and division. (Grades 1–6)	Math Media Division Cat.# M116	$1.90

Mathematics Using String—set of cards related to distances, thickness, graphing, open sentences and index of rotundity. Tasks are identified that require pupils to measure and record results. (Grades 3–6)	Math Media Division Cat.# M111	$1.00

Mirror Cards—The cards contain many colorful pictures and include four unbreakable mirrors. Pupils try to match a picture by using some combination of a	McGraw-Hill Book Co. Cat.# 07-018418-6	$12.95

card and its reflection. Cards are ordered by difficulty and test child's ability to make predictions and control patterns, as a basis for understanding more formal geometric concepts. 21 different sets of cards provide for a wide range of difficulty. Non-verbal components of cards make them attractive for children with learning difficulties. In later grades cards may be used to augment specific work in symmetry, optics, and geometry. Teacher's Guide available. (K-12)

Problem Cards—These cards provide graded sets of problems for Attribute Blocks, People Pieces and Color Cubes. Some of the cards are difficult for children to understand, but the problems are designed to help develop problem solving strategies and with proper teacher guidance can do so. Cards provide opportunity for problem solving using classification and relationships between classes. Many interesting games are either described or suggested in the cards. Teachers' Guide available. (K-12)

McGraw-Hill $9.60

Cat. #
07-07896-6

Student Activity Cards—126 activity cards that are coordinated with the Cuisenaire rods. Encourages individual or small group exploration with the rods into basic operations of addition, subtraction, multiplication and division, as well as fractions, factors, and multiples, and some geometrical relations. Cards are grouped by topic. Teacher's Guide available. (Grades 1–6)

Cuisenaire $9.95
Co. of
America, Inc.

Cat. # 20020

Tangram Cards—60 cards use seven piece dissection of the square to duplicate a variety of shapes. Divided into three levels of complexity and color coded. Teacher's Guide available. (K-12)

McGraw-Hill $4.62
Book Co.

Cat. #
07-017445-8

Triangle Cards—over 70 activity cards, together with some supportive material, provide the bases for many mathematical games and puzzles. Cards are best with small groups. Pattern recognition can be developed with the cards and related activities. The teacher's guide is included. (Grades K-12)

Selective $14.50
Educational
Equipment
Inc.

Cat. # XKIT02

Applied Mathematics Cards—150 cards present tasks requiring mathematics ideas and/or skills for pupils to complete. Some topics included are measurement, spatial relationships, plane and solid shapes and graphs. Many of the problem solving investigations can be accomplished outside of school. Cards are arranged in five groups of 30 cards. Each group represents a different level of difficulty. This material requires considerable facility in following highly verbal written directions. Teacher's guide available for each group. (Grades 3–9)

Math Media $34.00 Division

Cat. #M513

A Cloudburst of Math Lab Experiments— Over 900 experiments related to eight areas: 1) fundamental operations; 2) sets, numeration and number theory; 3) fractions; 4) probability and statistics; 5) geometry; 6) applications; 7) measurement; and 8) enrichment. Each activity is coded by topic and complexity, and any physical apparatus needed is identified. Activities can easily be integrated into most classrooms. Experiments are available in book or card format. There are four volumes of experiments, Elementary, Upper, Elementary, Junior High and Senior High School. Teacher's guide available. (K-12)

Midwest Publications, Inc.

	Book	Card
E Vol. 1	$2.95	$15.00
UE Vol. 2	3.95	16.00
JH Vol. 3	3.95	22.00
HS Vol. 4	2.95	12.00

Developmental Math Cards—12 packages of cards, lettered A through L, providing activities and open ended tasks related to five areas: 1) number; 2) measurement; 3) geometry; 4) games; and 5) notation. Difficulty levels increase between packages rather than within. Each package contains from 20 to 22 cards, with two packages suggested for each grade level. Teaching suggestions included. (Grades 1–6)

Addison-Wesley Publishing Co., Inc.

Cat#0540 Primary Set A-F $23.76

Cat.#0541 Junior Set G-L $23.76

Event Cards—100 cards that identify learning activities, experiments or games related to five mathematical areas: 1) number and operations; 2) geometry; 3) measurement; 4) statistics and probability; and 5) functions and graphs. Problem complexity varies within topics and the difficulty of each card is clearly identified.

Holt, Rinehart $49.95 and Winston Co.

Cat. # 03-086668-5

Physical quality of cards is excellent. Problems are well illustrated with many visual clues for the less facile reader. Card format is refreshing and highly motivating. (Grades 4–9).

Math Applications Kit—The 270 activity cards contain mathematical problems, experiments or questions related to five areas: 1) science; 2) social studies; 3) games and sports; 4) everyday things; and 5) occupations. Problem difficulty increases within each group of cards. Activities are designed to involve pupils in solving relevant and socially interesting problems. Teacher's guide available. (Grades 4–9)

Science Research $54.50
Associates Inc. (SRA)

Cat. #3–545

Math Activity Cards—240 plastic laminated cards that actively involve pupils in five basic areas: 1) graphs; 2) shapes; 3) measurements; 4) patterns; and 5) reasoning. Each of these areas are divided into five difficulty levels from A through E. Topic difficulty is between, not within levels. Cards stress concrete operations in open ended learning experiences. Teacher's guide available. (Grades 2–9)

Macmillan

Cat. #	Level	Price
28593	A	$12.60
28609	B	12.60
28627	C	12.60
28645	D	12.60
28662	E	12.60

Math Set Cards—two sets of cards providing a series of activity assignments dealing with measurement and geometry. Approximately 150 cards are ordered by difficulty for each topic. Designed for use individually or in small groups the cards are programmed sequentially providing for continuous progress. Cards have many written directions and would be appropriate for more advanced students at each grade level. Teacher's guide available. (Grades 3–8)

Scott Foresman
and Company

Cat. #02160
Geometry $11.25
Cat.# 02161
Measurement $11.25

No Read Math Activities—198 cards suggesting mathematical activities that can be completed in the classroom. These activities relate to the same eight areas as the Cloudburst Experiments. These activities are designed for poor readers. Most activities are illustrated pictorially and provide for a wide range of difficulty. Instructions to teacher on back of each card. (K-12)

Midwest Publications
Inc. $29.00
No Read Math
Activities

Open Ended Task Cards—18 concisely worded cards that challenge pupils to devise their own method of solving a meaningful mathematical problem. Activities involve pupils in measuring, counting, graphing, etc. No sequence is suggested for these open ended activities. Cards available in color or black and white. (Grades 4–9)

Teacher's Exchange of San Francisco

$5.00 color
2.00 b/w

Problem Cards—141 cards presenting a wide variety of mathematical problems. Ideal for individual and small group exploration. Problem difficulty level is noted by card color: 54-card Green set is easiest; 50-card Purple set next; and 37-card Red set most difficult. Teacher's guide available. (Grades 5–12)

John Wiley & Sons, Inc.

Cat. #	Set	Price
MO-471-65208-3	Green	$3.50
MO-471-65183-4	Purple	3.50
MO-471-65184-2	Red	3.50

Project Mathematics Activity Kit—84 cards related to the areas of geometry, measurement and number. Cards are ordered by difficulty and well illustrated, with many visual clues available for pupils with reading difficulties. Although the teachers's guide is coded for use with *Project Mathematics*, the activities provide practice and review of basic skills in the three above mentioned areas. (Grades K-3)

Mine Publications, Inc. $27.00

Cat. #0-03-921208-4

Willbrook Discovery Mathematics Workcards—174 activity cards dealing with four measurement topics: 1) time; 2) length; 3) weight; and 4) capacity. The cards are indexed according to these four topics and each card presents a task that requires pupils to do something. (Grades 1–6)

Selective Educational Equipment, Inc. (SEE)

Cat. #ES9889 $35.90

Workjobs—over 100 meaningful independent activities related to language development and mathematics. Each activity presents methods for getting started and ideas for follow up discussion. Activities are in a book format and accompanied by photographs. All necessary materials are readily available or can be reproduced by the teacher. Activities provide opportunities for pupils to develop skill in organizing thinking and expressing their thoughts. (Grades K-2)

Addison-Wesley Publishing Co., Inc.

Cat. #4311 $6.95

APPENDIX D

A GUIDE TO CURRENT MATHEMATICS TEACHING AIDS*

This Appendix attempts to organize the abundance of mathematics teaching aids available today. It lists materials found in catalogs from over 150 different companies.

The materials are classified under 24 general categories, which are listed below and denoted by capital letters. An explanation of materials found in each category is included under the category's title, whenever the title itself is not self-explanatory.

Table I summarizes the 24 general categories and identifies companies that produce materials classified within each of these categories. Companies are listed alphabetically and then numbered consecutively on the pages following Table I. The listing provides the complete mailing address for each company. It also includes a series of capital letters following each company which correspond to the categories and identify other teaching aids the company produces.

An attempt has been made to separate the materials into those appropriate for the elementary level (grades 1–6) and those appropriate for the secondary level (grades 7–12). Materials designated strictly for the elementary level will have subscript "e" following the category letter. Materials designated strictly for the secondary level will have subscript "s" following the category letter. Because there is considerable overlap, few categories are designated by either "e" or "s".

This guide helps you locate companies that produce materials in specific areas. *It does not provide an annotated list of commercial materials available*. If you wish specific information regarding specific products you may request a current catalog from the appropriate company. This is the only way of obtaining up to date information about prices, new materials, etc.

---

* Joan Kirkpatrick (University of Alberta) and Robert Jackson (University of Minnesota) developed the categorization scheme used in this appendix. A special thanks to Nancy English, Parkway School District, St. Louis, Missouri, for her help in preparing this guide.

CATEGORIES:

A. Colored Rods, Blocks, Beads and Discs (Also includes pattern blocks and attribute blocks)

B. Manipulative devices for the teaching of counting

C. Manipulative devices for the teaching of place value.

D. Manipulative devices for the teaching of mathematical operations and fractions. (Also includes devices for working with percent and decimals.)

E. Number boards
(Also includes geo-boards, array boards, and peg boards)

F. Cards and Instructional Kits
(Includes flash cards, activity cards, self-contained instructional kits, and self-contained mathematics labs)

G. Charts and Posters
(Also mobiles, manipulative charts, bulletin board materials, etc.)

H. Measurement Devices
(Includes tools for linear measurement, materials for measurement of area and volume, constructional devices, e.g. compass, etc.)

I. Models of Geometric Relationship
(Includes plane figures, solid figures, conic sections, polyhedra, trig models, problems dealing with geometric relationships, and some "make your own" kits)

J. Puzzles and Games

K. Film loops and strips

L. Films

M. Transparencies and Slides

N. Records and Tapes

O. Textbooks

P. Resource materials for the teacher
(Includes resource handbooks, booklets, pamphlets, manuals, films, professional books)

Q. Workbooks, duplicating masters

R. Enrichment and supplementary materials for the classroom.
(This is a broad category and deals primarily with software.
It identifies supplementary materials that can be used for either remedial or enrichment activities. Most of these materials were listed in Appendices A, B & C.)

S. Programmed Instruction
(Includes any sort of materials that are set up for programmed

learning. The materials include texts, workbooks, film cartridges, cards, etc.)

T. Calculating machines and computational devices (Includes slide rules, trig devices, tables, hand calculators, desk calculators, computers)

U. Chalkboards and chalkboard aids

V. Templates, stencils, plastic and wooden symbols, rubber stamps

W. Furniture and A-V Equipment

X. Materials written for CAI (Computer Assisted Instruction)

TABLE 1. This table summarizes the 24 categories of Mathematics Teaching Aids and identifies companies that distribute the materials

General Categories of Mathematics Teaching Aids		Commercial Distributor*
A	Colored Rods, Blocks, Beads and Discs	9, 20, 27, 29, 36, 39, 42, 48, 50, 56, 58, 59, 64, 65, 70, 78, 87, 88, 91, 93, 99, 103, 104, 106, 109, 116, 119, 121, 122, 127
B	Manipulative Devices for the teaching of counting	9, 13, 20, 23, 24, 27, 29, 38, 39, 42, 44, 48, 50, 53, 56, 58, 59, 64, 65, 68, 70, 75, 76, 78, 79, 84, 85, 87, 88, 89, 91, 93, 99, 101, 103, 104, 105, 106, 107, 109, 116, 119, 122, 124, 127, 131, 136, 139, 145
C	Manipulative Devices for the teaching of place value	9, 20, 23, 27, 29, 38, 42, 48, 50, 56, 58, 59, 64, 65, 70, 75, 78, 85, 88, 89, 91, 96, 99, 103, 104, 105, 106, 107, 109, 115, 116, 122, 127, 131, 145
D	Manipulative Devices for the teaching of mathematical operations and fractions	3, 6, 7, 9, 13, 16, 19, 20, 23, 24, 27, 29, 32, 38, 39, 42, 43, 44, 48, 50, 53, 58, 59, 62, 64, 65, 70, 73, 75, 76, 78, 79, 84, 85, 88, 89, 91, 93, 96, 97, 99, 101, 103, 104, 105, 106, 107, 109, 115, 116, 118, 119, 120, 122, 124, 127, 131, 133, 139, 145, 146
E	Number boards	9, 20, 25, 27, 29, 42, 48, 50, 58, 59, 64, 65, 70, 75, 78, 85, 87, 88, 91, 99, 103, 104, 106, 107, 109, 116, 118, 121, 122, 127, 142, 145, 152
F	Cards and Instructional kits — F	11, 15, 25, 48, 59, 64, 65, 76, 85, 87, 91, 96, 97, 104, 109, 113, 118, 122, 125, 127, 132, 142, 146
	F_e	3, 9, 20, 29, 32, 38, 42, 49, 50, 53, 58, 70, 75, 78, 79, 88, 94, 99, 100, 103, 106, 107, 116, 120, 121, 130, 133, 145, 147
	F_s	36, 66, 83, 84, 137

* Numerals instead of companies are listed. The company name may be found by checking the appropriate numerals on pages 280–288.

G	Charts and Posters	G	20, 39, 44, 45, 47, 48, 52, 59, 64, 76, 103, 104, 105, 107, 109, 122, 127, 140
		G_e	9, 42, 50, 56, 58, 70, 88, 99, 116, 131
		G_s	51, 55, 84, 129, 141
H	Measurement Devices	H	9, 20, 28, 38, 42, 48, 53, 56, 58, 59, 64, 65, 70, 76, 80, 83, 84, 85, 88, 91, 93, 95, 99, 101, 103, 104, 105, 107, 109, 111, 122, 127
		H_e	27, 50, 75, 121
		H_s	135, 152
I	Models of Geometric Relationships	I	4, 9, 15, 20, 23, 27, 29, 36, 38, 40, 42, 44, 48, 53, 54, 55, 57, 58, 59, 64, 65, 70, 84, 88, 91, 92, 93, 101, 103, 104, 109, 116, 119, 122, 126, 127, 139, 146
		I_e	7, 50, 78, 79, 85, 99, 106, 120
		I_s	17, 83, 152
J	Puzzles and Games	J	2, 9, 11, 15, 18, 23, 25, 26, 29, 35, 36, 45, 48, 58, 59, 62, 64, 69, 70, 73, 81, 82, 85, 88, 91, 93, 96, 97, 99, 101, 103, 104, 107, 109, 110, 113, 116, 118, 120, 122, 124, 127, 132, 133, 141, 146, 148, 151
		J_e	19, 20, 22, 27, 38, 42, 49, 50, 68, 78, 79, 102, 106, 112, 114, 128, 136
		J_s	55, 83
K	Film loops and strips	K	20, 23, 37, 40, 44, 56, 84, 104, 115, 112, 124, 146
		K_e	48, 103
		K_s	33, 93
L	Films	L	5, 40, 46, 72, 74, 77, 86, 105, 108, 115, 134, 144, 146, 150
		L_e	29
		L_s	51, 93, 140
M	Transparencies and slides	M	9, 14, 20, 23, 48, 56, 58, 59, 60, 61, 62, 64, 65, 75, 76, 92, 98, 99, 101, 103, 104, 106, 109, 118, 119, 120, 121, 127, 139
		M_e	42, 50, 70
		M_s	33, 80, 83, 84, 93
N	Records and Tapes	N	60, 65, 87, 94, 106, 109, 118, 121
		N_e	9, 16, 19, 20, 48, 92, 104, 116
O	Textbooks	O	1, 35, 40, 56, 60, 62, 65, 93, 94, 100, 106, 109, 113, 118, 120, 130, 133
		O_e	29
		O_s	21, 63, 149
P	Resource materials for the teacher	P	1, 27, 35, 38, 41, 45, 55, 62, 65, 70, 76, 84, 85, 93, 96, 99, 100, 103, 104, 105, 106, 113, 120, 130, 132, 141, 142, 149
		P_e	20, 48, 50, 58, 79, 116, 133
		P_s	21, 63

Q	Workbooks, duplicating masters	Q	9, 23, 38, 45, 48, 56, 58, 60, 61, 64, 65, 70, 88, 93, 94, 99, 103, 106, 109, 113, 118, 119, 120, 121, 127, 130, 133, 142, 146
		Q_e	20, 32, 50, 98, 116, 147
		Q_s	141
R	Enrichment and supplementary materials for the class room	R	1, 8, 14, 15, 29, 35, 36, 40, 41, 42, 45, 52, 56, 58, 60, 62, 65, 84, 85, 87, 91, 93, 94, 96, 99, 100, 104, 105, 106, 113, 120, 122, 127, 132, 138, 141, 146, 149,
		R_e	79, 116, 147
		R_s	2, 21, 55, 63
S	Programmed Instruction	S	40, 60, 62, 65, 89, 93, 106, 115, 146
		S_e	12, 48
		S_s	55, 92, 135, 149
T	Calculating machines and computational devices	T	6, 9, 11, 24, 28, 29, 30, 34, 36, 48, 58, 59, 64, 71, 83, 84, 88, 99, 101, 103, 104, 105, 109, 111, 122, 125, 127, 135, 139, 143, 152
		T_s	17, 80, 90
U	Chalkboards and chalkboard aids	U	9, 20, 27, 34, 38, 48, 53, 58, 59, 64, 67, 70, 84, 88, 99, 103, 104, 109, 116, 122, 126, 127, 145
		U_e	39, 50, 75, 78
		U_s	55, 83, 92, 135, 152
V	Templates, stencils, plastic and wooden symbols, rubber stamps		9, 20, 28, 36, 42, 45, 48, 50, 53, 55, 58, 59, 64, 75, 78, 80, 84, 88, 97, 99, 103, 104, 109, 111, 116, 126, 127, 135, 145
W	Furniture and A-V Equipment		9, 10, 20, 23, 27, 28, 31, 42, 48, 55, 58, 59, 70, 80, 84, 88, 92, 94, 99, 101, 103, 104, 109, 115, 119, 123, 127, 146
X	Materials written for CAI (Computer Assisted Instruction)		41, 118, 120

NAMES AND ADDRESSES OF 152 COMPANIES THAT DISTRIBUTE MATHEMATICS TEACHING AIDS

1. Addison-Wesley Publishing
   Co., Inc.
   Reading, Massachusetts
   01867

   O, P, R

2. The Advancement Placement
   Institute
   169 North 9th Street
   Brooklyn, New York 11211

   J, R_s

3. Aero Educational Products
   St. Charles, Illinois 60174

   D, F_e

4. Aestheometry, Inc.
   1903 Clayton Ave.
   Artesia, New Mexico 88210

   I

5. American Bankers Association
   Public Relations Department
   90 Park Avenue
   New York, New York 10018

   L

6. Arithmetical Principles
   Association
   5848 N.E. 42nd Avenue
   Portland, Oregon 97218

   D, T

7. Auto-Instructional Materials
   AIM Industries, Inc.
   253 State Street
   St. Paul, Minnesota 55107

   D, I_e

8. Automobile Manufacturers
   Association Educational
   Services
   320 New Center Building
   Detroit, Michigan 48202

   R

9. Beckley-Cardy Company
   1900 North Harragansett Ave.
   Chicago, Illinois 60639

   A, B, C, D, E, F_e, G_e, H, I,
   J, M, N_e, Q, T, U, V, W

10. Bell and Howell
    Audio-Visual Sales Dept.
    7100 McCormick Road
    Chicago, Illinois 60645

    W

11. Bell Telephone Laboratories
    Mountain Avenue
    Room 3B-228
    Murray Hill, New Jersey 07974

    F, J, T

12. Benefic Press
    10300 W. Roosevelt Road
    Westchester, Illinois 60153

    S_e

13. Ben-G-Products, Inc.
    462 Sagamore Ave.
    Williston, New York 11596

    B, D

14. Channing L. Bete Company,
    Inc.
    45 Federal Street
    Greenfield, Massachusetts
    01301

    M, R

15. Book-Lab Inc.
    Dept. AT1
    1449 – 37th Street
    Brooklyn, New York 11218

    F, I, J, R

16. Bremner Multiplication
    Records, Inc.
    161 Green Bay Road
    Wilmette, Illinois 60091

    D, N_e

17. Brooks Manufacturing
Company (Trig-Aide)
P.O. Box 41195
Cincinnatti, Ohio 45241

I_s, T_s

18. Cadaco, Inc.
310 West Polk Street
Chicago, Illinois 60607

J

19. Caddy-Imler Creations, Inc.
Box 5097
Inglewood, California 90310

D, J_e, N_e

20. Cambosco Scientific
Company, Inc.
342 Western Avenue
Boston, Massachusetts 02135

$A, B, C, D, E, F_e, G, H, I$
$J_e, K, M, N_e, P_e, Q_e, U, V, W$

21. Cambridge University Press
510 North Avenue
New Rochelle, New York
10801

O_s, P_s, R_s

22. Childcraft Education Corp
964 Third Avenue
New York, New York 10022

J_e

23. Colonial Film and Equipment
Co., Inc.
752 Spring Street, N.W.
Atlanta, Georgia 30308

B, C, D, I, J, K, M, Q, W

24. Comspace Corporation
350 Great Neck Road
Farmingdale, L.I., New York
11735

B, D, T

25. Concept Company
P.O. Box 273
Belmont, Massachusetts 02178

E, F, J

26. Cooperative Recreation
Service, Inc.
Radnor Road
Delaware, Ohio 43015

J

27. Creative Publications
P.O. Box 10328
Palo Alto, California 94303

$A, B, C, D, E, H_e, I, J_e, P,$
U, W

28. The C-Thru Ruler Company
6 Britton Drive
Bloomfield, Connecticut
06002

H, T, V

29. Cuisenaire Company of
America, Inc.
12 Church Street
New Rochelle, New York 10805

$A, B, C, D, E, F_e, I, J, L_e, O_e, R, T,$

30. The Curta Company
P.O. Box 3414
Van Nuys, California 91405

T

31. Custom Fabricators, Inc.
8702 Bessemer Avenue
Cleveland, Ohio 44127

W, also Portable Math Lab

32. Dana and Company, Inc.
P.O. Box 201
Barrington, Rhode Island
02806

D, F_e, Q_e

33. Denoyer-Geppert
5235 Ravenswood Avenue
Chicago, Illinois 60640

K_s, M_s

34. Digiac
Division of Digital Electronics,
Inc.
Ames Court
Plainview, Long Island,
New York
11803

T

35. Dover Publications
180 Varick Street
New York, New York 10014

J, O, P, R

36. Edmund Scientific Company
101 Gloucester Pike
Barrington, New Jersey 08007

A, F_s, I, J, R, T, V

37. Educational Development
Laboratories (EDL)
Huntington, New York 11743

K

38. Educational Research and
Development, Inc.
Box 466
Green Lake, Wisconsin 54941

B, C, D, F_e, H, I, J_e, P, Q, U

39. EduKaid of Ridgewood
1250 East Ridgewood Avenue
Ridgewood, New Jersey 07450

A, B, D, G, U_e

40. Encyclopedia Britannica
Instructional Materials Division
151 Bloor Street West
Toronto 5, Ontario

I, K, L, O, R, S

41. Entelek Incorporated
42 Pleasant Street
Newburyport, Massachusetts
01950

P, R, X

42. ETA School Materials Division
159 West Kinzie Street
Chicago, Illinois, 60610

A, B, C, D, E, F_e, G_e, H, I,
J_e, M_e, R, V, W

43. Exton Aids
Box MT
Millbrook, New York 12545

D

44. Eye-Gate House
146-01 Archer Ave.
Jamaica, New York 11435

B, D, G, I, K

45. Fearon Publishers
2165 Park Boulevard
Palo Alto, California 94306

G, J, P, O, R, V

46. Films Incorporated
1144 Wilmette Ave.
Wilmette, Illinois 60091

L

47. Ford Motor Company
Educational Affairs Dept.
The American Road
Dearborn, Michigan 48121

G

48. Gamco Industries, Inc.
P.O. Box 1911 A
Big Spring, Texas 79720

A, B, C, D, E, F, G, H, I, J, K_e,
M, N_e, P_e, Q, S_e, T, U, V, W

49. Garrard Publishing Company
(Dolch)
123 West Park Avenue
Champaign, Illinois 61820

F_e, J_e,

50. Gel-Sten Supply Company,
Inc.
911–913 South Hill Street
Los Angeles, California 90015

A, B, C, D, E, F_e, G_e, H_e,
J_e, M_e, P_e, Q_e, U_e

51. General Electric Company
Educational Publishers
450 Duane Avenue
Schenectady, New York 12301

G_s, L_s

52. General Motors Corp.
Public Relations Staff
3044 W. Grand Blvd.
Detroit, Michigan 48202

G, R,

53. Genius Supply Company
7288 N. Teutonia Ave.
Milwaukee, Wisconsin 53209

B, D, F_e, H, I, U, V

54. Geodestix
P.O. Box 5179
Spokane, Washington 99205

I

55. Geyer Instructional Aids
Company
404 East Hawthorne Street
Fort Wayne, Indiana 46806

G_s, I, J_s, P, R_s, S_s, U_s, V, W

56. Ginn and Co.
450 West Algonquin Road
Arlington Heights, Illinois
60005

A, B, C, G_e, H, K, M, O, Q, R

57. A. J. Gude 3rd
845 Dudley
Lakewood, Colorado 80215

I

58. G. W. School Supply Specialists
P.O. Box 14
Fresno, California 93707

A, B, C, D, E, F_e, G_e, H, I, J, M,
P_e, Q, R, T, U, V, W

59. J. L. Hammett Company
Hammett Place
Braintree, Massachusetts
02184

A, B, C, D, E, F, G, H, I, J, M,
T, U, V, W

60. Harcourt, Brace Jovanovich,
Inc.
7555 Caldwell Avenue
Chicago, Illinois 60648

M, N, O, Q, R, S

61. Hayes School Publishing Co.,
Inc.
321 Penwood Avenue
Wilkinsburg, Pennsylvania
15221

M, Q

62. D. C. Heath and Co.
2700 North Richardt Avenue
Indianapolis, Indiana 46219

D, J, M, O, P, R. S

63. Holt, Rinehart and
Winston Co.
383 Madison Avenue
New York, New York 10017

O_s, P_s, R_s

64. Jack Hood School
Supplies Co., Ltd.
91–99 Erie Street
Stratford, Ontario

A, B, C, D, E, F, G, H, I, J, M,
Q, T, U, V

65. Houghton Mifflin Company
Dept. K
110 Tremont Street
Boston, Massachusetts 02107

A, B, C, D, E, F, H, I, M, N,
O, P, Q, R, S

66. Howell Enterprises, Ltd.
P.O. Box 176
Littlerock, California 93543

F_s

67. Hubbard Scientific Company
P.O. Box 105
Northbrook, Illinois 60062

U

68. Hudson Products
5120 Colby, Apt. 2
Everett, Washington 98201

B, J_e

69. IBM
Armonk, New York 10504

J

70. Ideal School Supply Company
11000 South Lavergne Avenue
Oaklawn, Illinois 60453

A, B, C, D, E, F_e, G_e, H, I, J,
M_e, P, Q, U, W

71. Illinois Tool Works, Inc.
Cutting Tool Division
2501 N. Keeler Avenue
Chicago, Illinois 60639

T

72. University of Illinois
Visual Aids Service
Division of University Extension
Champaign, Illinois 61820

L

73. Imout (Educational Drill
Games)
706 Williamson Bldg.
Cleveland, Ohio 44114

D, J

74. Indiana University
Audio-Visual Center
Bloomington, Indiana 47401

L

75. The Instructo Corporation
Paoli, Pennsylvania 19301

B, C, D, E, F_e, H_e, M, U_e, V

76. The Instructor Publications,
Inc.
Dansville, New York 14437

B, D, F, G, H, M, P

77. International Film Bureau
332 S. Michigan Ave. (MT)
Chicago, Illinois 60604

L

78. The Judy Company
310 North 2nd Street
Minneapolis, Minnesota
55401

A, B, C, D, E, F_e, I_e, J_e, U_e, V

79. Kenworthy Educational
Services
P.O. Box 3031
138 Allen Street
Buffalo, New York 14205

B, D, F_e, I_e, J_e, P_e, R_e

80. Keuffel and Esser Company
2507 Jefferson St.
Kansas City, Missouri 64108

H, M_s, T_s, V, W

81. Kohner Brothers, Inc.
Tryne Game Division
P.O. Box 294
East Paterson, New Jersey
07407

J

82. James W. Lang
P.O. Box 224
Mound, Minnesota 55364

J

83. Lano Company
4741 West Liberty Street
Ann Arbor, Michigan 48103

F_s, H, I_s, J_s, M_s, T, U_s

84. LaPine Scientific Company
6001 South Knox Avenue
Chicago, Illinois 60629

B, D, F_s, G_s, H, I, K, M_s, P, R,
T, U, V, W

85. Learning Research Associates,
Inc.
1501 Broadway
New York, New York 10036

B, C, D, E, F, H, I_e, P, R

86. Madison Project
918 Irving Ave.
Syracuse, New York 13210

L

87. Mafex Associates, Inc.
111 Barron Ave. Box 519
Johnston, Pennsylvania 15907

A, B, E, F, N, R

88. Mainco School Supply
Company
57 Pine Street
Canton, Massachusetts 02021

A, B, C, D, E, F_e, G_e, H, I, J,
Q, T, U, V, W

89. Mast Development Company
Dept. MT-8
2212 E. 12th St.
Davenport, Iowa 52803

B, C, D, S

90. Math – Aide
Sturbridge Office Center
Sturbridge, Massachusetts
01566

T_s

91. Math Media Division
H and M Associates
P.O. Box 1107
Danbury, Connecticut 06810

A, B, C, D, E, F, H, I, J, R

92. Math-U-Mathic, Inc.
3017 North Stiles
Oklahoma City, Oklahoma
73105

I, M, N_e, S_s, U_s, W

93. McGraw-Hill Book Company
Educational Games and Aids
330 West 42nd Street
New York, New York 10036

A, B, D, H, I, J, K_s, L_s, M_s, O,
P, Q, R, S

94. Charles E. Merrill Books, Inc.
1300 Alum Creek Drive
Columbus, Ohio 43216

F_e, N, O, Q, R, W

95. Metric-Association, Inc.
2004 Ash Street
Waukegan, Illinois 60085

H

96. Midwest Publications Co., Inc.
P.O. Box 129
Troy , Michigan 48084

C, D, F, J, P, R

97. Miles Kimball Company
41 West Eighth Avenue
Oshkosh, Wisconsin 54901

D, F, J, V

98. Milliken Publishing Company
611 Olive Street Road
St. Louis, Missouri 63101

M, Q_e

99. Milton Bradley Company
School Department
P.O. Box 1581
Springfield, Massachusetts
01101

A, B, C, D, E, F_e, G_e, H, I_e, J,
M, P, Q, R, T, U, V, W

100. Mine Publications, Inc.
25 Groveland Terrace
Minneapolis, Minnesota
55403

F, P, O, R

101. Minnesota Mining and
Manufacturing 3M Company
3M Center
St. Paul, Minnesota 55101

B, D, H, I, J, M, T, W

102. Motivational Research, Inc.
4216 Howard Avenue
Kensington, Maryland 20795

J

103. Moyer Division
Vilas Industries Limited
25 Milvan Drive
Weston, Ontario, Canada

A, B, C, D, E, F_e, G, H, I, J, K_e,
M, P, Q, T, U, V, W

104. Nasco Mathematics
Fort Atkinson, Wisconsin
53538

A, B, C, D, E, F, G, H, I, J,
K, M, N_e, P, R, T, U, V, W

105. National Council of Teachers
of Mathematics (NCTM)
1201 Sixteenth Street N.W.
Washington, D.C. 20036

B, C, D, G, H, L, P, R, T

106. Thomas Nelson and Sons
81 Curlew Drive
Don Mills, Ontario

A, B, C, D, E, F_e, I_e, J_e, M,
N, O, P, Q, R, S

107. Nifty Division
St. Regis Paper Company
3300 Pinson Valley Parkway
P.O. Box 6416
Birmingham, Alabama 35217

B, C, D, E, F_e, G, H, J

108. Official Films Inc.
776 Grand Ave.
Ridgefield, New Jersey 07657

L

109. Palfreys School Supply
Company
7715 East Farvey Boulevard
Rosemead, California 91770

A, B, C, D, E, F, G, H, I, J, M,
N, O, Q, T, U, V, W

110. Parker Brothers
Box 900
Salem Massachusetts 01970

J

111. Pickett Inc.
Pickett Square
436 E. Gutierrez St.
P.O. Box 1515
Santa Barbara, California
93102

H, T, V

112. Playskool Manufacturing Co.
3720 North Kedzie Avenue
Chicago, Illinois 60618

J_e

113. Prindle, Weber & Schmidt,
Inc.
53 State Street,
Boston, Massachusetts 02109

F, J, O, P, Q, R,

114. Responsive Environmental
Corporation
Learning Materials Division
Englewood Cliffs, New Jersey
07632

J_e

115. Sargent-Welch Scientific Co.
7300 North Linder Avenue
Skokie, Illinois 60076

C, D, K, L, S, W

116. School Service Company
647 South LaBrea Avenue
Los Angeles, California 90036

A, B, C, D, E, F_e, G_e, I, J, N_e,
P_e, Q_e, R_e, U, V

117. Schoolhouse Visuals, Inc.
816 Thayer Avenue
Silver Springs, Maryland 20910

J_e

118. Science Research Associates,
Inc.
259 East Erie Street
Chicago, Illinois 60611

D, E, F, J, M, N, O, Q, X

119. Scott Education Division
Scott Graphics, Inc.
Holyoke, Massachusetts 01040
Includes: Jam Handy, EML
and Technifax

A, B, D, I, M, Q, W

120. Scott, Foresman and
Company
1900 East Lake Avenue
Glenview, Illinois 60625

D, F_e, I_e, J, M, O, P, Q, R, X

121. Scott Scientific, Inc.
P.O. Box 2121
Fort Collins, Colorado 80521

A, E, F_e, H_e, M, N, Q

122. Selective Educational
Equipment (SEE) Inc.
3 Bridge Street
Newton, Massachusetts
02195

A, B, C, D, E, F, G, H, I, J, K,
R, T, U

123. E. H. Sheldon Equipment
Company
Muskegon, Michigan

W

124. Singer
Education and Training
Products
Society for Visual Education,
Inc.
1345 Diversay Parkway
Chicago, Illinois 60614

B, D, J, K

125. Southwestern Bell Telephone
Co.
301 S. Jefferson
Springfield, Missouri 65806

F, T

126. The Speed-Up Geometry
Ruler Co., Inc.
14 Osborne Avenue
Baltimore, Maryland 21228

I, U, V

127. St. Paul Book & Stationery Co.
1233 West Co. Rd. E.
St. Paul, Minnesota 55112

A, B, C, D, E, F, G, H, I, J, M,
Q, R, T, U, V, W

128. Stanley Bowman Company,
Inc.
4 Broadway
Valhalla, New York 10595

J_e

129. L. S. Starett Company
Athol, Massachusetts 01331

G_s

130. Steck-Vaughan Company
P.O. Box 2028
Austin, Texas 78767

F_e, O, P, Q

131. Teacher's Aids
1683 South 700 West
Woods Cross, Utah 84087

B, C, D, G_e

132. Teachers' Exchange of
San Francisco
600 35th. Avenue
San Francisco, California
94121

F, J, P, R

133. Teachers Publishing
Corporation (Publishers of
Grade Teacher)
23 Leroy Avenue
Darien, Connecticut 06820

D, J, O, P_e, Q

134. Teaching Film Custodians
25 W. 43rd St.
New York, New York 10036

L

135. Teledyne Frederick Post
P.O. Box 803
Chicago, Illinois 60690

H_s, S_s, T, U_s, V

136. Touch, Inc.
P.O. Box 1711
Albany, New York 12201

B, J_e

137. Fern Tripp
Special Teaching Materials for
Special Classes
2035 E. Sierra Way
Dinuba, California 93618

F_s

138. Charles E. Tuttle
Company, Inc.
P.O. Box 470
Rutland, Vermont 05701

R

139. Tweedy Transparencies
208 Hollywood Avenue
East Orange, New Jersey 07018

B, D, I, M, T

140. U.S. Dept. of Health,
Education and Welfare
National Audiovisual Center
U.S. Government Films
Washington, D.C. 20409

L_s

141. J. Weston Walch, Publisher
Box 658 Main Post Office
Portland, Maine 04104

G_s, J, P, Q_s, R

142. Walker Educational Book
Corp.
720 Fifth Avenue
New York, New York 10019

E, F, P, Q

143. Wang Laboratories, Inc.
Dept. 12 BY
836 North Street
Tewksbury, Massachusetts
01876

T

144. Wayne University
Audio-Visual Materials
438 W. Ferry St.
Detroit, Michigan 48202

L

145. Weber Costello
1900 North Narragansett
Avenue
Chicago, Illinois 60639

B, C, D, E, F_e, G, U, V

146. Webster Division
McGraw-Hill Book Company
Manchester Road
Manchester, Missouri 63011

D, F, I, J, K, L, Q, R, S, W

147. Western Publishing
Company, Inc.
1220 Mound Avenue
Racine, Wisconsin 53404

F_e, Q_e, R_e

148. WFF'N Proof
Box 71
New Haven, Connecticut
06501

J

149. John Wiley and Sons, Inc.
605 Third Avenue
New York, New York 10016

O_s, P, R, S_s

150. University of Wisconsin
Bureau of Visual Instruction
P.O. Box 2093
Madison, Wisconsin 53701
(Limited to Wisconsin and
bordering states)

L

151. World Wide Games
Box 450
Delaware, Ohio 43105

J

152. Yoder Instruments
East Palastine, Ohio 44413

E, H_s, I_s, T, U_s

APPENDIX E

THE MATHEMATICS LABORATORY IN HISTORICAL CONTEXT

The growth and development of mathematics laboratories has been influenced by many factors. Figure 1 provides a brief overview of the movement and identifies some of the influential agents.

OVERVIEW OF THE GROWTH AND DEVELOPMENT OF MATHEMATICS LABORATORIES

Stage	Influential Agents
Early Development (Pre–1890)	Frederick Froebel, Heinrich Pestalozzi, Jean Rousseau
Sowing the Seed (1890–1920)	John Dewey, Karl Goldziher, A. R. Hornbrook, George W. Myers, E. H. Moore, Frances W. Parker, John Perry, J. W. Young
Germination Period (1921–1955)	Commission on Post-War Plans, National Council of Teachers of Mathematics
Initial Growth (1956–1970)	COLAMBA, Madison Project, Minnemath Project, NCTM, Nuffield Mathematics Project, UICSM
Current Status (Post 1970)	Pupils, teachers, schools, teacher education programs, professional organizations and commercial publishers.

Figure 1

This figure does not tell the complete story, but it presents an interesting perspective of the evolution of mathematics laboratories. It should also remind us that today's "mathematics laboratories" are not new, but the result of seeds planted long ago and nurtured by many people having diverse philosophies.

The idea of learning by doing is as old as civilization itself. This concept of learning, however, was not evident in formal school programs for centuries. The teacher, lecturer, or tutor was looked upon as the primary source of knowledge. The teacher was the purveyor of truth, the omnipotent being in the classroom. Consequently, the format for the transmission of knowledge was primarily expository from teacher to student.

During the 18th and 19th centuries, several progressives, such as Fredrick Froebel, Heinrich Pestalozzi, and Jean Rousseau advocated drastic changes in the educational structure of the schools. Among other things they contended that children should be treated as children, not miniature adults, and that the natural curiosity of youngsters should be put to good advantage in the schools. As a result, they emphasized the importance of allowing children to learn through experience. The exclusive use of formal textbook instruction was considered unnatural, at least until the children observed a specific need for such experiences. Furthermore, they suggested that teachers use means of instruction other than verbalization and recommended providing a variety of sensory experiences for children. In spite of their outspoken criticism of existing educational practices, their demands for pedagogical change fell, for the most part, upon deaf ears.

In the decade from 1890–1900, John Perry advocated change and succeeded in basing mathematics instruction in England's technical schools on experimental and laboratory work. He said that a change away from a pure abstract approach to teaching mathematics was necessary and in its place, the use of graphs, models, physical apparatus and practical applications was needed. In an effort to summarize the basic elements of the Perry Movement the following principles were formulated by Alderson:

1. The necessity for breaking down the artificial barriers between different subjects and parts of subject.

2. The fact that for the great mass of our students, utility is of greater importance than philosophic speculation.

3. The fact that mathematics is the language of science and must be taught as such.

4. The fact that much of what we now prove laboriously can be assumed and time saved for interesting and original work.

5. Physics and mathematics must be taught together.[1]

Many of these points have long been advocated and in some cases practiced by mathematics teachers. The astute reader will observe

---

[1] Victor A. Alderson, "Five Cardinal Points in the Perry Movement," *School Mathematics*, 4:193, March 1904.

historical parallels between the Perry Movement and the contemporary emphasis on active learning as well as the correlation of science and mathematics in today's curricula.

The Perry Movement in England had a pronounced influence upon the reorganization of secondary school mathematics in the United States. More specifically, Perry provided the rationale and impetus needed to generate some degree of interest in the development of methodology which facilitates the effective uses of instructional materials in teaching mathematics.

It seems appropriate to recall a postscript to the Perry Movement reported in a 1912 article written by a proponent of the movement, indicating that there has been a lot of talk but very little action.[2] Although it would be unwise to overdramatize the effectiveness of the Perry Movement, much of the current experimental work in mathematics education is indirectly related to the works of John Perry.

During the same period, proponents of developmental child psychology also began to advocate the idea of teaching children as children. Men such as G. Stanley Hall were able to make their views on child psychology known during the early part of the 20th century. Although this active period of developmental psychology produced some changes, most notably in curriculum sequencing and methodology, it failed to gain widespread acceptance in educational circles. The learning atmosphere in most classrooms was still characterized by teacher dominated activities. As a result most students in both elementary and secondary schools assumed a passive role in the educative process.

A review of the literature in mathematics education prior to 1950, reveals that several articles were published that discussed active involvement of students in learning mathematics.* The publications during this period directed their attention toward the physical components of a mathematics laboratory with heavy emphasis placed on exercises which typified the utility role of mathematics. Consequently the scope of these reports was quite limited, with little attention given to either sharing exemplary classroom lessons or discussions of a wide variety of learning activities appropriate for different levels.

During the decade of the fifties, professional journals contained more articles relating to mathematics laboratories than had been published prior to 1950. Perhaps the most widely read were the articles by Emil Berger in THE MATHEMATICS TEACHER from October, 1950, through June, 1954. These articles were concerned with the practical implications which instructional devices have for the teaching of mathematics. Although many ingenious plans were presented, the focus was frequently on instructional aids to be used exclusively by the teacher. As a result student involvement

---

2 Joseph V. Collins, "The Perry Idea in the Mathematical Curriculum," *School Science and Mathematics*, 12:296–299, April 1912.

* Articles and related publications are identified in the bibliography.

with these instructional materials was quite limited. Despite the fact that much was written during the 1950s, this decade ended with relatively few mathematics laboratories in existence.

Recent Conditions Inhibiting the Development of Mathematics Laboratories

Several factors tended to inhibit the growth of mathematics laboratories during the fifties and early sixties. The first was a lack of funds to equip and support the mathematics laboratory. The National Defense Education Act (NDEA) of 1958 represented a significant effort by the Federal Government to support educational innovation. A portion of these funds was earmarked for inservice education of teachers, while a substantial percentage was to be used for acquisition of instructional materials. During this same period the National Science Foundation (NSF) began to financially support mathematics and science curriculum development as well as institutes for mathematics and science teachers throughout the country. The financial support of NDEA and NSF has had a significant and positive effect on developments, inherent in the revolution of school mathematics. Their effects, however, have been far more within the realm of general curriculum development than within the specific realm of promoting the use of the mathematics laboratories.

A second factor impeding the growth of the materials laboratory was the mathematics education of teachers. The early NSF institutes and teacher workshops were content oriented, and perhaps rightly so, since the mathematical competency of the average teacher was in need of upgrading. The primary objective of most inservice education programs was thus to strengthen the mathematics background of the participants. As a result, little or no attention was given to the consideration of a variety of instructional strategies in light of contemporary theories of mathematics learning.

Substantial efforts have not been made to prepare mathematics teachers for their classroom role in the promotion of active student involvement in the learning of mathematics. Certainly different instructional methods at all levels are necessary if the stereotyped image of the mathematics teacher is to be removed. An examination of teacher education programs reveals that far too few programs prepare prospective elementary and/or secondary mathematics teachers to assume a classroom role in a laboratory setting.

Lack of adequate curriculum materials is a third factor which has slowed the growth and progress of mathematics laboratories. Much needs to be done in the broad realm of curriculum development. For example, the domain consisting of laboratory lessons and assignment cards is practically untouched. Many teachers operating in a laboratory setting have found it impossible to obtain commercially prepared laboratory lessons appropriate for *their* classroom needs. This situation has prompted teachers, both individually and in groups, to write their own assignment cards for laboratory use.

Other teachers have chosen to modify the commercially produced assignment cards and use these revisions in the classroom. Both of these plans should eventually lead to an increase in the quantity and quality of clear assignments for use in the mathematics classroom. The horizon looks increasingly promising in this respect. Commercial publishers are currently realizing the untapped potential in the area of the materials laboratory, and are directing their efforts accordingly.

Another crucial area in curriculum development which has hampered the development of the mathematics laboratory has been the need for continuity in the sequence of learning activities. The development of laboratory lessons about selected mathematics topics has produced a piecemeal program. Unfortunately, the non-trivial responsibility of integrating and coordinating these laboratory lessons into the regular mathematics program has been, for the most part, left to the classroom teacher. The lack of continuity in laboratory activities has discouraged many teachers from effectively utilizing them. Although efforts have been made to improve this articulation in the mathematics curriculum, much work remains.

One of the first formal programs in the United States which advocated active learning of mathematics in an informal setting was the Madison Project. Under the able leadership of Robert Davis, this project has made substantial contributions relative to the creation of activities and the devising of instructional schema designed to change the medium of learning. The Madison Project Shoebox series was one of the Madison Project's most publicized and best-received products. This series served as a catalyst for the continuing development and use of the host of independent study activities available to today's classroom teacher.

England's Nuffield Project has provided a model for large scale implementation of an active approach to the learning of mathematics. Nuffield's "the world is my classroom" philosophy is reflected in their curriculum materials and in the teacher inservice component, so essential to the success of their program. Teacher workshops acquaint teachers with the materials, provide opportunity for teachers to exchange ideas and in general prepare them to assume their new role in mathematics classrooms.

Although the limiting factors of money, teacher education, and curriculum have not by any means been overcome, the growth of mathematics laboratories during the 1960s was phenomenal. Few, if any, reports on the number of mathematics laboratories in existence were made prior to 1960. In 1968 Phillips reported that some 2,000 laboratories were operating in elementary, junior, and senior high schools.[3] Research reporting recent changes in junior and senior high school programs identified the rapid increase in the number of mathematics laboratories as the most noticeable cur-

---

[3]   Phillips, *op. cit.*, p. 1.

ricular trend from 1965 to 1968.[4] The authors, aware of the difficulties in identifying a mathematics laboratory, recognize the possible error in such research. Nevertheless, this does suggest an observable trend toward the continued growth of active learning in a mathematics laboratory setting.

BIBLIOGRAPHY—APPENDIX E

Alderson, Victor A. "Five Cardinal Points in the Perry Movement." *School Mathematics*, 1904.

Alspaugh, John; R. D. Kerr; and Robert E. Reys, "Curriculum Change in Secondary School Mathematics." *School Science and Mathematics*, 70:(1970):170–176.

Collins, Joseph V. "The Perry Idea in the Mathematical Curriculum." *School Science and Mathematics*. 12:(1912):296–299.

Davis, Robert B. "The Changing Curriculum in Mathematics." Washington, D.C.: Association for Supervision and Curriculum Development, 1967.

Emerging Practices in Mathematics Education, 22nd Yearbook. Washington, D.C.: National Council of Teachers of Mathematics, 1954.

Goldziher, Karl. "Mathematical Laboratories." *School Science and Mathematics*, 8:(1908):753–757.

Hall, G. Stanley. *Aspects of Childlife and Education*. Boston: Ginn and Co., 1921.

History of Mathematics Education in United States and Canada, 32nd Yearbook. Washington, D.C.: National Council of Teachers of Mathematics, 1970.

Kidd, Kenneth P.; Shirley S. Myers; and David M. Cilley. *The Laboratory Approach to Mathematics*. Chicago: Science Research Associates, 1970.

Multi-Sensory Aids in the Teaching of Mathematics, 18th Yearbook. Washington, D.C.: National Council of Teachers of Mathematics, 1945.

Phillips, Harry L. "The Mathematics Laboratory." *American Education*, 1:(1968):1–3.

Potter, Mary A. "The Mathematics Laboratory." *School Science and Mathematics*, XLIV: 367–373, April, 1944.

Sims, Weldon and Albert Oliver. "The Laboratory Approach to Mathematics." *School Science and Mathematics*, L: 621–627, November, 1950.

---

[4] John Alspaugh, R. D. Kerr and Robert E. Reys, "Curriculum Change in Secondary School Mathematics," *School Science and Mathematics*, 70:174, February 1970.

INDEX